VFD Challenges for Shipboard Electrical
Power System Design

VFD Challenges for Shipboard Electrical Power System Design

Mohammed M. Islam

IEEE PRESS

WILEY

Published by John Wiley & Sons, Inc., Hoboken, New Jersey.
Published simultaneously in Canada.

For general information on our other products and services or for technical support, please contact our Customer Care Department within the United States at (800) 762-2974, outside the United States at (317) 572-3993 or fax (317) 572-4002.

Wiley also publishes its books in a variety of electronic formats. Some content that appears in print may not be available in electronic formats. For more information about Wiley products, visit our web site at www.wiley.com.

Library of Congress Cataloging-in-Publication Data is available.

ISBN: 9781119463382

Set in 10/12pt Warnock by SPi Global, Pondicherry, India

Printed in the United States of America.

V10015006_102719

Dedicated to my loving wife, Raihana Islam,
and my wonderful children, Tanjila Islam and Nasheet Islam

Contents

Preface

The variable frequency drive (VFD) motor controllers are considered better selection over other types of motor controllers for shipboard applications. The VFDs are also used as adjustable speed drive (ASD) because of simple operational capabilities. The adjustable speed drives are suitable for ship electrical propulsion system. The adjustable speed drive electrical propulsion is excellent choice for the cruise ship due to less vibration comparing with direct drive propulsion system.

The shipboard VFD application along with system level ungrounded power generation plays a vital role for the system integrity. The shipboard low voltage ungrounded power system ground detection and monitoring system is usually a simple detection system with lights for monitoring voltage variations. IEC has developed completely different recommendations from the ground detection light with the understanding that the legacy system does not contribute to the management of real grounding danger as the system leads to arcing and then bolted fault. The IEC requirement is to monitor and intervene as the electrical system starts making transition from symmetric to asymmetric behavior. Additionally, any ground in the ungrounded electric system on ships, corrective action must be fast enough to protect the system from arcing fault, explosion, and then related equipment failures.

In relation with the VFD and ASD application challenges, the following electrical system fundamentals will be based upon:

1) 6600 V power generation and distribution system is specially for high power variable frequency drive and adjustable speed drive such as propulsion and propulsion related thrusters, etc.
2) 690 V power generation and distribution system mainly for dedicated bus system to isolate the VFD motor controllers from 480 V ship service distribution system. This arrangement is mainly to isolate the VFD related electric noise to ensure the 480 V ship service bus is protected.
3) Ungrounded low voltage ship service power generation and distribution such as 480 V, three-phase three-wire system and 120 V, three-phase

three-wire distribution system. *(Remarks: There are VFD applications when the low voltage power distribution is a four-wire system with a neutral wire connected to the ground. This type of ground path creates unwanted EMI related circulating current throughout the systems, power, control, and signals with an adverse effect on equipment and systems. In addition, if the four-wire system is not fully balanced then the unbalanced current flows through the ground path and circulates with the VFD generated electrical noise. Therefore, for VFD and ASD populated shipboard power systems the four-wire distribution system is not recommended. IEEE-45 discourages the general use of low voltage, 208 V/120 V, three-phase, four-wire system on ships).*

4) N+1 Power generation requirements
5) Power system black out recovery
6) Motor controllers, across the line, soft start, reduced voltage starting
7) VFD Motor controller
8) Understanding VFD application, harmonic calculation and harmonic management
9) Harmonic filter application
10) Special identification of VFD/ASD cable as VFD cable
11) VFD cable selection, routing, segregation, and termination
12) Electrical system grounding and ground detection

The selection of cable for adjustable speed drive application is also a fundamental challenge. This book provides in depth analysis of the cable application challenges with recommended solutions.

The fundamentals for "all electric ship" power generation and distribution are the major consumer for the "variable frequency drives (VFD)" for propulsion system and many other VFD consumer loads. The VFD provides many operational advantages such as easy control of the propeller RPM and delivering required torque and very low speed. Similarly, the VFD also brings many other technical issues which the design engineer must understand and take appropriate measures so that challenging issues are engineered properly. Some of the challenging issues are:

a) Electrical noise such as harmonic, including high frequency noise
b) Understanding of VFD drive application, harmonic generation, harmonic calculation, harmonic management, special cable requirements, VFD cable installation and termination requirements
c) Grounding matters at the ungrounded generation level
d) Grounding matters at the ungrounded distribution level
e) Grounding matters at the low voltage distribution
f) Grounding matters at the equipment level
g) Single point grounding matters for MV and LV systems
h) Medium voltage system protection and coordination
i) Failure mode and its effect analysis (FMEA)

The terminologies used for the design and development of VFD/ASD related equipment are mainly by IEC standard. IEC terminologies and symbols are different from the ANSI terminology and symbols. It is very important to understand the difference between IEC and ANSI standard electrical devices.

Author introduced Ship Smart System Design (S3D) concept which is physics-based simulation and virtual prototyping of overall ship design and then compare with the real system of system interaction for electrical power generation, distribution, protection and automation. The S3D can be better visualized by major transformation of industrial automation revolution, "Industry 4.0" with low cost computing, powerful software analytical platforms of data mining, machine learning and data networking technologies. The S3D is embodied in "Industry 4.0", "Digital Twin" which will definitely enhance the life cycle of ship design, development, operation, and maintenance.

There are many electrical elementary one-line diagrams presented for experienced engineers who will be able to analyze different aspects of shipboard electrical distribution systems and then select the most appropriate one for application. If any one of the electrical elementary one-line diagrams falls beyond the regulatory body requirements, appropriate adjustment must be made to ensure full compliance with regulations.

I am grateful to the initial reviewers for recommending to add AC/DC/AC, AC/DC, DC/DC, DC/AC application details. Chapter 3 has been added for the LVDC and MVDC distribution along with VFD/ASD at the distribution with many different combinations of user interface.

I realize that the design engineers may not be involved in IEEE-45 and related standard development process. I recommend the design engineers stay within the boundary of IEEE-45 recommendations for the use of VFD/ASD applications, VFD cable type selection, VFD cable installation and VFD related harmonics.

Mohammed M. Islam

About the Author

Mohammed M. (Moni) Islam is R&D Manager of Applied Science at Northrop Grumman Ship Systems. He has 45 years of diversified shipboard electrical engineering experience and has played significant roles in every part of new shipbuilding and ship modernization engineering. Mr. Islam also currently serves as the IEEE-45 central committee Vice-Chair and is a member of IEEE-1580 working group. He has been involved in the "All Electric Ship" R&D programs for many years and was the principal investigator of the ship Smart-System Design (S3D) feasibility study, an ONR funded research and development project. He received his Bachelor of Marine Engineering Technology from the Merchant Marine Academy of Bangladesh in 1969, and Bachelor of Electrical Engineering Degree with Honors from the State University of New York, Fort Schuyler Maritime College, in 1975.

1

Overview – VFD Motor Controller

The variable frequency drive (VFD) motor controller has been proven to be suitable for shipboard low voltage and medium voltage application. The variable speed electric propulsion is very popular due to the full range of speed and torque controls. The medium voltage variable speed electric propulsion can achieve high propeller power a with full range of speed controls.

The VFD motor controller produces electrical noise. The standards recommend a wide range total harmonic distortion (THD) with limits of from 5% to 8%. The author recommends the shipboard power system THD should not be more than 5%. This 5% THD is allowed when all other design attributes are fully complied with.

There are situations when the THD can be up to 8%, but that distribution system must be isolated from the other system so that the electric noise is managed within the dedicated system.

The 5% harmonic contents can create a dangerous situation affecting other low voltage equipment and control circuits. The design engineers must understand the THD limits for the specific design and should consider these as a system level harmonic management approach.

It is highly recommended that one manufacturer VFD be used and the manufacturer guidelines for harmonic management, VFD cable selection and VFD cable termination should be followed. Otherwise the coordination with multiple VFD suppliers can be challenging.

The following requirements will be discussed as to the use of VFD:

1) The IEEE-45-2002 recommends an acceptable THD limit of 5%, which is in line with MIL-STD-1399. The IEEE-45-2002 also allows 8% THD only for a dedicated bus with dedicated VFD equipment suitable for 8% harmonic environment.
2) IEEE-519-1992 recommends an acceptable THD limit of 5%.

VFD Challenges for Shipboard Electrical Power System Design, First Edition.
Mohammed M. Islam.
© 2020 by The Institute of Electrical and Electronics Engineers, Inc.
Published 2020 by John Wiley & Sons, Inc.

3) IEEE-519-2014 recommends an acceptable THD limit of 8% with clarification that the IEEE-519 may not apply to the shipboard power system.
4) The IEEE-45.1-2017 recommends an acceptable THD limit of 8% for normal bus and dedicated bus.
5) ABS Steel Vessel rules 2017 have a THD limit of 8%.

For simplicity, voltage source inverters (VSI) will be presented. The VSI are fed with ship service constant voltage. The VFD electronics convert the AC to DC and then DC to AC voltage with adjustable magnitude and frequency.

The VSI drives use capacitive storage, with capacitors in their DC link, which both stores and smooths the DC voltage for the inverter input.

There are types of power switching devices used with variable frequency drives. We will discuss the use of IGBT which are semiconductor switches that are turned on and off, creating a pulse width modulated (PWM) output with regulated frequency.

For all shipboard ungrounded power generation and distribution systems, there is a possibility of a system level ground path, such as HRG, filter. These ground loops should be minimized and regulated, so that VFD generated noise such as common mode voltage and current can travel through the low resistance ground path.

System level capacitance monitoring and management is also recommended.

If there are HRGs used in an ungrounded system, total probable current path shall be monitored and managed to establish and maintain safe level.

All high frequency harmonics must be calculated and managed to establish a safe harmonic level. This may require harmonic calculation beyond the 49th harmonic of IEEE-519 requirements. The IEEE-519-2014 may not be applicable for establishing the shipboard harmonic level at the low frequency input site of the point of common coupling (PCC) due to the lack of ground reference in an ungrounded system. The IEEE-519 may not address the high frequency harmonics generated by the inverter side of the VFD. Therefore, the shipboard VFD application and inverter side high frequency noise must be managed, so that the high frequency EMI and RFI generated equipment malfunctional can be avoided.

In view of the high frequency noise propagation all over the shipboard power system, the cable between the VFD motor controller and the motor must be as recommended by the VFD manufacturer. The VFD cable usually has three-phase conductors, three drain wires and overall shield. The VFD cable must be terminated with drain wires and shielding, as recommended by the cable manufacturer and the VFD manufacturer. For additional details of VFD cable requirements and termination requirements refer to Chapter 6.

1.1   MIL-STD-1399 Shipboard Power System and Total Harmonic Requirements

The shipboard 60Hz power system is outlined in MIL-STD-1399-300 as follows:

a) The ship service electrical power distribution system supplied by the ship's generators is 440 Vrms, 60 Hz, three-phase, ungrounded.
b) Power for the ship's lighting system and other user equipment such as electronic equipment supplied from the ship power distribution system through transformers, is 115 Vrms, 60 Hz, three-phase, ungrounded.

Remarks: The MIL-STD-1399 does not address ungrounded 440 Vrms or 115 Vrms power generation and distribution. For an ungrounded distribution system (440 Vrms or 115 Vrms), engineering analysis should be performed to minimize the neutral current circulation in the distribution system by balancing single-phase and three-phase actual loads (See Table 1.1).

1.2   Shipboard Power System Design Fundamentals

Electrical power system detailed design and development for commercial ships, such as cruise ships, cargo ships, tankers, related support vessels, offshore industry related floating platforms, and all other support vessels is featured in this book. Some military ship designs are also included to establish differences in design fundamentals as to the requirements of redundancy requirements and zonal distributions. The design requirements and fundamentals are with the understanding of the following:

a) Regulatory requirements
b) Operational requirements
c) Redundancy requirements
d) Understanding of emergency requirements as to the power generation as well as the emergency load distribution.
e) Understanding the causes of blackout (Dead ship) situation. The blackout situation for all electric ship related power generation and distribution is more complex than the ship with non-electric propulsion.
f) Electric propulsion related power generation and distribution requirements have been taken in the design to adapt medium voltage power generation. Due to the fact that ample power is available to change the hydraulic system or mechanical system to electric system with variable drive operation.

Table 1.1 MIL-STD-1399 Shipboard Power System Characteristics.

MIL-STD-1399-300B

TABLE I. <u>Characteristics of shipboard electric power systems.</u> (NOTE: Characteristic percentages are defined in Section 3.)

Characteristics	Type I	Type II	Type III
Frequency			
1) Nominal frequency	60 Hz	400 Hz	400 Hz
2) Frequency tolerance	±3% (±5% for submarines)	±5%	±0.5%
3) Frequency modulation	0.5%	0.5%	0.5%
4) Frequency transient tolerance	±4%	±4%	±1%
5) Worst case frequency excursion from nominal resulting from items 2, 3, and 4 combined, except under emergency conditions	±5.5%	±6.5%	±1.5%
6) Recovery time from items 4 or 5	2 seconds	2 seconds	0.25 second
Voltage			
7) Nominal user voltage	440, 115, 115/200 Vrms	440, 115 Vrms	440, 115, 115/200 Vrms
8) Line-to-line voltage unbalance	3% (0.5% for 440 Vrms, 1% for 115 Vrms for submarines)	3%	2%
9) User voltage tolerance			
a) Average line-to-line voltage from nominal	±5%	±5%	±2%
b) Line-to-line voltage from nominal, including items 8 and 9a	±7%	±7%	±3%
10) Voltage modulation	2%	2%	1%
11) Maximum departure voltage from nominal resulting from items 8, 9a, 9b, and 10 combined, under transient or emergency conditions	±8%	±8%	±4%
12) Voltage transient tolerance	±10%	±10%	±5%
13) Worst case voltage excursion from nominal resulting from items 8, 9a, 9b, 10 and 12 combined, except tinder emergency conditions	±20%	±20%	±5.5%
14) Recovery time from item 12 or item 13	2 seconds	2 seconds	0.25 second
15) Voltage spike (± peak value)	2.5 kV (440 Vrms sys) 1.0 kV (115 Vrms sys)	2.5 kV (440 Vrms sys) 1.0 kV (115 Vrms sys)	2.5 kV (440 Vrms sys) 1.0 kV (115 Vrms sys)

g) The grounding requirements are different than the traditional low voltage distribution though both systems are three-wire ungrounded systems.

h) Vital auxiliary must be properly classified. There are regulatory requirements of vital auxiliary related redundant services and operational requirements which directly contributes to the design and development.

In general, shipboard electrical system is ungrounded. The ungrounded system has no dedicated neutral line in the distribution system. However, there is always a capacitive ground path. This phenomenon of capacitive ground path needs to be better understood in view of system level grounding and bonding.

For better understanding, the grounding and bonding will be called G and the neutral line will be called N.

The non-linear solid state power applications usually create rapid changes to voltage and current while transferring energy to the load. These changes cause high frequency current to flow to the ground. This is considered as electrical noise.

There are many good features of electric drive (VFD and ASD) related applications onboard ships and platforms. However, there are many features which may contribute as electric noise such as harmonics, transients, grounding at the equipment level and system level. The design engineer must understand those issues, so that the causes and effects are properly analyzed during concept design and detail design. Recent VFD-related failure reports warrant better understanding, better design, and then overall design management. Electrical propulsion and auxiliary service requirements for the use of VFD contributed to recent operational challenges due to critical operational issues.

There are sample designs in support with electrical one line (EOL) diagrams are presented in this book to explain operational requirements which are unique for each class of vessels leading to a customize design. The designs are presented with drawings and diagrams, so fundamental electrical design steps are discussed, and then compared with the service requirements. This includes ship service power requirements, power requirements, and emergency power requirements.

The design includes:

1) Shipboard electrical low voltage and medium voltage power generation, electrical propulsion and power distribution systems. The fundamental shipboard electrical design requirements, design details, verification of the design prior to equipment installation, and then verification of the test results to establish a design base for the ship.

2) Offshore floating platforms and offshore support vessels as applicable.

The shipboard power system ground detection system is provided to detect ground in the ungrounded system so that the ground lifted as soon as the ground occurs. Single-phase ground fault is detectable; however, the system will continue to operate on the other two healthy phases. However, a second ground fault, phase to phase will create arcing, which must be monitored and lifted as soon as possible. Then for a three-phase fault, which is also called bolted fault, which must be detected as fast as possible, then the protective system must isolate the bolted fault to avoid any kind of explosion:

i) The solid-state devices operate with some ground reference. The basic requirements of a shipboard ungrounded system may not be complied with.
ii) The resistance grounding system use delta-wye transformer with wye neutral connected to ground with resistor is also an established ground reference point in the ungrounded power system. The resistance grounded system and ungrounded low voltage distribution system create a ground loop in the entire power system.
iii) The ungrounded power system ground plane in ideal conditions is a zero-voltage reference point in ideal conditions only.
iv) The delta-wye configuration is not acceptable for shipboard installations as it can propagate electric noise with the wye distribution. The delta–delta configuration is recommended as both primary and secondary will help to circulate electric noise within the winding.
v) The delta–wye will circulate noise all over the distribution system due to the wye configuration. Again, it is very difficult to maintain zero ground reference in a wye distribution system. The grounded wye distribution system creates a ground plan coupled with an ungrounded zero reference point.

Remarks: There are some special cases where the grounded wye distribution system is allowed due to operator safety reasons, such as an electrical workshop where the operator may use hand-held electrical tools. These features will not be discussed to avoid any misrepresentation (See Figures 1.1 to 1.5).

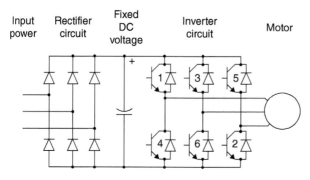

Figure 1.1 Typical VFD Power Circuit Topology With 6 Pulse Diode Rectifier, DC Link Filter and Voltage Source Inverter Showing Common Mode Voltage and Current.

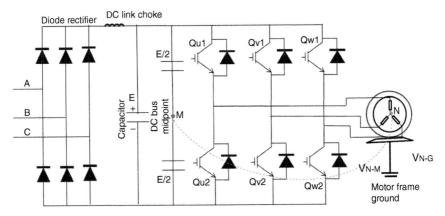

Figure 1.2 Typical VFD Power Circuit Topology With 6 Pulse Diode Rectifier, DC Link Filter, and Voltage Source Inverter Showing Common Mode Voltage and Current.

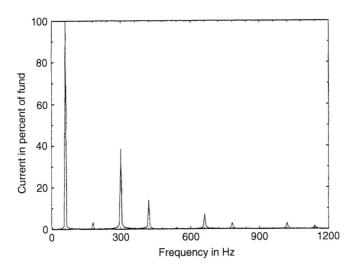

Figure 1.3 PWM Drive Harmonic Spectrum (Six Pulse Drive).

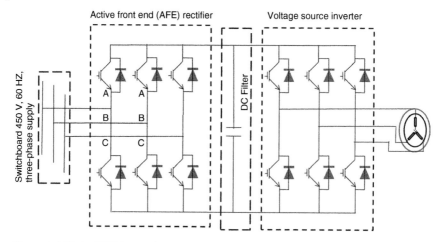

Figure 1.4 Typical VFD Power Circuit Topology With PWM Active Front End (AFE) DC Filter and Voltage Source Inverter.

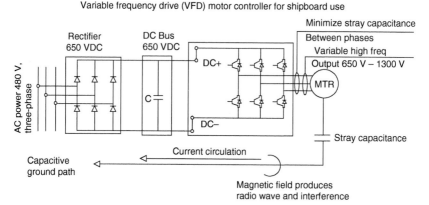

Notes:
1. Do not anchor the VFD driven motor winding neutral point to the ground.
2. If the main supply line is grounded, the capacitive ground will circulate throughout the system ground and will contaminate the entire system. Shipboard power system is usually ungrounded, the shipboard ungrounded power supply will also be effected throughout through stray ground path.
3. Minimize stray system capacitance. Minimize stray capacitance between cable conductors.
4. Use VFD rated cable with proper voltage rating, drain wire and shielding.
5. Use short length VFD cable as recommended by the VFD supplier.

Figure 1.5 Typical VFD Power Circuit Topology With 6 Pulse Diode Rectifier, DC Link Filter, and Voltage Source. Inverter Showing Common Mode Voltage and Current.

1.3   Shipboard Low Voltage Power System Design Development With VFD and Verification

Refer to Figures 1.6 to 1.8 and Table 1.2.

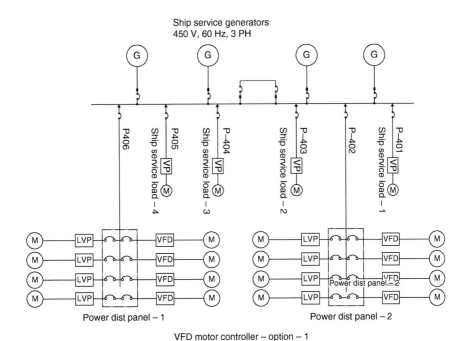

Figure 1.6 Typical EOL With Ship Service and Emergency Generator.

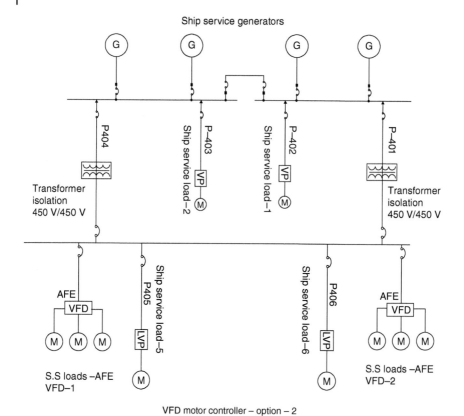

Figure 1.7 Typical EOL With Dedicated VFD Bus.

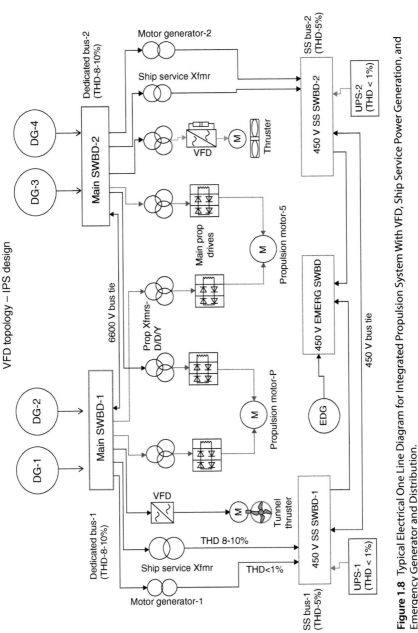

Figure 1.8 Typical Electrical One Line Diagram for Integrated Propulsion System With VFD, Ship Service Power Generation, and Emergency Generator and Distribution.

Table 1.2 Typical VFD Common Mode Voltage and Current.

COMMON MODE VOLTAGE IDENTIFICATION	
Balanced Three-phase System Motor Input	**VFD Driven Three-phase Output – Motor Input**
$V_{N\text{-}G}$ is zero	$V_{N\text{-}M}$ may not be zero (Common mode voltage)
	$V_{N\text{-}G}$ may not be zero (Common mode voltage)

Remarks: Icom is the common mode current flowing through the cable surface due to common mode voltage. This common mode current must be drained to the ground by a ground path such as the drain wire, cable shield, etc. Therefore, the VFD cable is identified as cable including drain wire and cable shield with higher insulation withstand capabilities.

1.4 Low Voltage Motor and Cable Insulation Stress Due to Variable Frequency Drive

The VFD output may is much higher than the input voltage. Per IEEE-45 the VFD motor cable voltage shall be 3.5 times the input voltage as shown in the Table 1.3.

Table 1.3 Low Voltage Motor and Cable Insulation Stress.

VAC RMS Voltage-Input	VAC Peak Voltage	Rectified VDC = VAC_{PK}	Inverted VAC_{PK}	VAC RMS Voltage-INV-Output	Voltage Swing-INV Out-Peak-Peak	VFD Cable Insulation
Rectifier Input-$VAC_{PK(RMS\text{-}IN)}$	Rectifier Input-VAC_{PK}	VDC = VAC_{PK}	VDC = VAC_{PK} (Inverted)	$VDC = VAC_{RMS-OUT}$ $= \dfrac{Vac\ pk}{\sqrt{2}}$		
120 $V_{RMS\text{-}IN}$	170 VAC $_{PK}$	170 VDC	340 VAC $_{PK}$	170 VDC	340 VAC $_{PP}$	600 V/1 kV
208 $V_{RMS\text{-}IN}$	294 VAC $_{PK}$	294 VDC	558 VAC $_{PK}$	294 VDC	558 VAC $_{PP}$	600 V/1 kV
460 $V_{RMS\text{-}IN}$	650 VAC $_{PK}$	650 VDC	1300 VAC $_{PK}$	650 VDC	1300 VAC $_{PP}$	2 kV
600 $V_{RMS\text{-}IN}$	849 VAC $_{PK}$	849 VDC	1698 VAC $_{PK}$	849 VDC	1698 VAC $_{PP}$	2 kV
690 $V_{RMS\text{-}IN}$	975 VAC $_{PK}$	975 VDC	1950 VAC $_{PK}$	975 VDC	1950 VAC $_{PP}$	2 kV
720 $V_{RMS\text{-}IN}$	1036 VAC $_{PK}$	1018 VDC	2036 VAC $_{PK}$	1018 VDC	2036 VAC $_{PP}$	2 kV
2000 $V_{RMS\text{-}IN}$	2828 VAC $_{PK}$	2828 VDC	5656 VAC $_{PK}$	2828 VDC	5656 VAC $_{PP}$	5 kV

1.5   Shipboard Power Quality and Harmonics Requirements

1.5.1   IEEE Std 45-2002, Clause 4.6, Power Quality and Harmonics

Solid state devices such as motor controllers, computers, copiers, printers, and video display terminals produce harmonic currents. These harmonic currents may cause additional heating in motors, transformers, and cables. The sizing of protective devices should consider the harmonic current component. Harmonic currents in nonsensically current waveforms may also cause EMI and RFI. EMI and RFI may result in interference with sensitive electronics equipment throughout the vessel.

Isolation, both physical and electrical, should be provided between electronic systems and power systems that supply large numbers of solid state devices, or significantly sized solid state motor controllers. Active or passive filters and shielded input isolation transformers should be used to minimize interference. Special care should be given to the application of isolation transformers or filtering, as the percentage of power consumed by solid state power devices compared with the system power available increases. Small units connected to large power systems exhibit less interference on the power source than do larger units connected to the same source. Solid state power devices of vastly different sizes should not share a common power circuit. Where kilowatt ratings differ by more than 5 to 1, the circuits should be isolated by a shielded distribution system transformer. Surge suppressers or filters should only be connected to power circuits on the secondary side of the equipment power input isolation transformers.

Notes:

1) To preclude radiated EMI, main power switchboards rated in excess of 1 kV and propulsion motor drives should not be installed in the same shipboard compartment as ship service switchboards or control consoles. (This is per IEEE 45-1998 Clause 4.6).

2) To reduce the effect of radiated EMI, special considerations on filtering and shielding should be exercised, when main power switchboards and propulsion motor drives are installed in the same shipboard compartment as ship service switchboards or the control console.

3) IEEE Std 519™-1992 provides additional recommendations regarding power quality. The IEEE-519-2014 is the latest edition which is widely different from the 1992 version. Reference to both standards will be necessary to establish the state of the requirement and apply as recommended.

1.5.2   Power Conversion Equipment Related Power Quality

1.5.2.1   IEEE Std 45-2002, Clause 31.8, Propulsion Power Conversion Equipment (Power Quality)

The following quote is an extract referring only to the power quality portion of this clause:

> Whenever power converters for propulsion are applied to integrated electric plants, the drive system should be designed to maintain and operate with the power quality of the electric plant. The effects of disturbances, both to the integrated power system and to other motor drive converters, should be regarded in the design. Attention should be paid to the power quality impact of the following:
>
> a) Multiple drives connected to the same main power system.
> b) Commutation reactance, which, if insufficient, may result in voltage distortion adversely affecting other power consumers on the distribution system. Unsuitable matching of the relation between the power generation system's sub-transient reactance and the propulsion drive commutation impedance may result in production of harmonic values beyond the power quality limits.
> c) Harmonic distortion can cause overheating of other elements of the distribution system and improper operation of other ship service power consumers.
> d) Adverse effects of voltage and frequency variations in regenerating mode.
> e) Conducted and radiated electromagnetic interference and the introduction of high-frequency noise to adjacent sensitive circuits and control devices. Special consideration should be given for the installation, filtering, and cabling to prevent electromagnetic interference.

1.6   Ship Smart System Design (S3D) for System Design With VFD (See Chapter 9)

The THD limits must be properly applied for specific distribution system to meet the operational requirements. For guide refer to Table 1.4.

Table 1.4 THD Limits for Various Applications.

Item	System	THD limits	Discussion
1	Dedicated bus-main switchboard	5 to 8%	Bus provides VFD power supply only
2	Sensitive bus- ship service switchboard	Less than 5%	Some critical equipment onboard
3	Non-Sensitive bus-ship service switchboard	Max 5%	
4	Emergency bus-Emergency switchboard	@ 1%	
5	UPS	Less than 1%	Guided by the system requirements

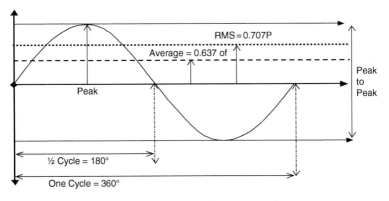

Figure 1.9 Typical EoI With Ship Service and Emergency Generator.

1.7 Carrier Frequency Ranges for Typical Solid State Devices

Refer to Table 1.5.

Table 1.5 Carrier Frequency for SCR, BJT and IGBT.

DEVICE TYPE	CARRIER FREQUENCY	EXPLANATION	REMARKS
SCR	250 Hz to 500 Hz	(a) 4 to 8 time the fundamentals	
BJT	I kHz to 2 kHz	(a) 8 to 16 times the fundamentals	
IGBT	2 kHZ to 20 kHz	(a) 16 to 160 times the fundamentals	

Solid state devices such as motor controllers, computers, copiers, printers, and video display terminals produce harmonic currents. These harmonic currents may cause additional heating in motors, transformers, and cables. The sizing of protective devices should consider the harmonic current component. Harmonic currents in nonsensically current waveforms may also cause EMI and RFI. EMI and RFI may result in interference with sensitive electronics equipment throughout the vessel.

Isolation, both physical and electrical, should be provided between electronic systems and power systems that supply large numbers of solid state devices, or significantly sized solid state motor controllers. Active or passive filters and shielded input isolation transformers should be used to minimize interference. Special care should be given to the application of isolation transformers or filtering as the percentage of power consumed by solid state power devices compared with the system power available increases. Small units connected to large power systems exhibit less interference on the power source than do larger units connected to the same source. Solid state power devices of vastly different sizes should not share a common power circuit. Where kilowatt ratings differ by more than 5 to 1, the circuits should be isolated by a shielded distribution system transformer. Surge suppressors or filters should only be connected to power circuits on the secondary side of the equipment power input isolation transformers.

To reduce the effects of radiated EMI, special considerations on filtering and shielding should be exercised when main power switchboards and propulsion motor drives are installed in the same shipboard compartment as ship service switchboards or control consoles.

1.8 VFD Fundamentals

See Figures 1.10 to 1.14.

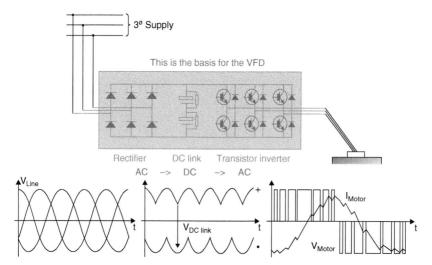

Figure 1.10 Frequency Spectrum for Input, Rectifier Output, and PWM Showing Uneven Current Output.

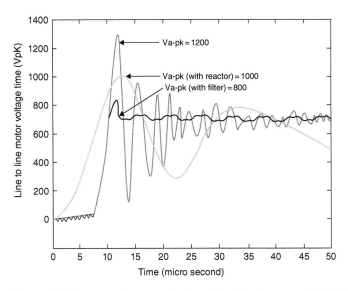

Figure 1.11 Frequency Spectrum for Input, Rectifier Output, and PWM Showing Uneven Current Output.

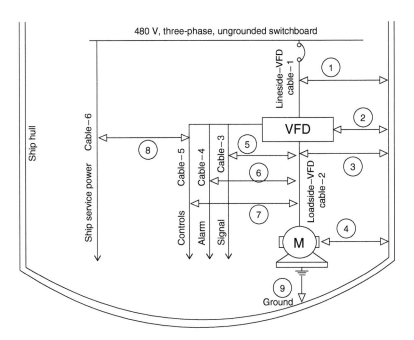

Figure 1.12 Recommended VFD and VFD Cable Separation.

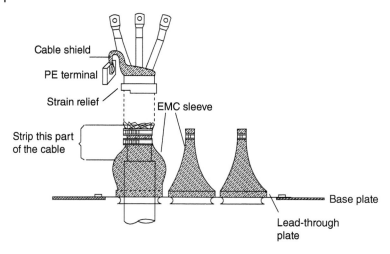

Cable shield

PE terminal

Strain relief

EMC sleeve

Strip this part of the cable

Base plate

Lead-through plate

Figure 1.13 VFD Cable Shield 360 Degree High Frequency Termination With EMC Sleeve.

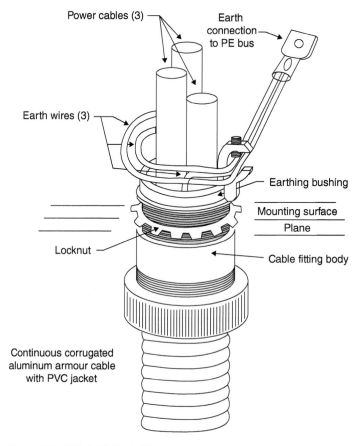

Power cables (3)

Earth connection to PE bus

Earth wires (3)

Earthing bushing

Mounting surface

Plane

Locknut

Cable fitting body

Continuous corrugated aluminum armour cable with PVC jacket

Figure 1.14 VFD Cable Drain Wire 360 Degree Termination.

1.9 Apparent Power for Linear and Non-linear Loads With and Without Harmonics

See Figures 1.15 and 1.16.

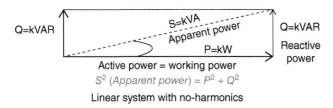

Linear system with no-harmonics

Figure 1.15 Power Vector for Linear System With No Harmonics.

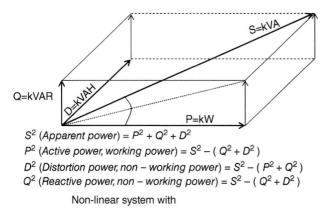

$$S^2 \text{ (Apparent power)} = P^2 + Q^2 + D^2$$
$$P^2 \text{ (Active power, working power)} = S^2 - (Q^2 + D^2)$$
$$D^2 \text{ (Distortion power, non - working power)} = S^2 - (P^2 + Q^2)$$
$$Q^2 \text{ (Reactive power, non - working power)} = S^2 - (Q^2 + D^2)$$

Non-linear system with

Figure 1.16 Power Vector for Non-Linear System with Harmonics.

1.10 Ship Smart System Design (S3D) – "Digital Twin"

The Smart ship system design (S3D) concept can be embodied in "digital twin" in view of low cost computing, powerful analytical platforms of data mining, machine learning, and data networking technology. Any complex ship design and development fidelity can be achieved by using "digital twin", so that there is no need for arbitrary simulation, costly hardware in the loop prototyping.

2

Propulsion System Adjustable Speed Drive

Traditionally, the ship propulsion system propeller directly driven by the prime mover or through gear box has many mechanical challenges such as a long drive shaft and shaft alignment, and complex gear box.

The VFD for propulsion systems has gained popularity due to drive technology maturity for the following:

i) The propulsion shaft need not be in line with the engine.
ii) The propulsion motor can be very close to the propeller, with total elimination of propulsion shaft and shaft alignments.
iii) Electrical speed and torque control of the propulsion.
iv) Electrical system smooth ramp-up and down of the ship speed.
v) Propulsion diesel engines running at a synchronous speed for generator.
vi) Eliminating propulsion engine noise due to varying engine speed.
vii) The diesel engine location is variable as DG can be in any convenient location.
viii) Ease of using solid state technology.

The shipboard power system design process has become very complex due to high power generation and variable drive. Traditional design experience is not sufficient to support the new challenges of high power generation, high power consumables, such as electrical propulsion, and other adjustable drive ship service applications. Traditional protection systems such as over load and short circuit protection is not sufficient. This book addresses both design and development issues namely:

a) System design process with VFD.
b) System developmental process with VFD.
c) Verification of the Design & Development with VFD for regulatory requirements.

VFD Challenges for Shipboard Electrical Power System Design, First Edition.
Mohammed M. Islam.
© 2020 by The Institute of Electrical and Electronics Engineers, Inc.
Published 2020 by John Wiley & Sons, Inc.

d) Verification of failure mode and effects of the VFD design.
e) Verification of training requirements for the VFD design.
f) Verification of operator readiness for VFD equipment.

2.1 The Shipboard Propulsion Power ASD/VFD Anomalies

See Table 2.1.

Table 2.1 VFD Influence on Shipboard Power System.

Item	Issues	Causes	Discussion
1	Premature motor failure	High frequency components of the inverter output creating electrical noise such as (a) Electromagnetic interference (EMI); Electromagnetic radiation and Electromagnetic induction (b) Circulation of common mode current, (c) Reflected wave	**Electromagnetic radiation** – Can damage equipment, and the drive Traditional three conductor cable will act as a large radiating loop antenna effecting shipboard electrical control system, radio devices and alarms. Specially designed power cable with shielding and drain wire will manage these problems, however, cable procurement, termination, and routing must be properly done. • Effective overall shielding will greatly reduce EMI. • Shield and ground wire should be terminated properly. • Ensure 360 degree termination of the shield **Electromagnetic Induction** – VFD induces magnetic field by rectifying AC to DC and then inverting DC to AC by the use of switching at a very high rate. Common mode Current – In ideal situation there should not be any common mode current. However, the common mode current can be the conductor current plus ground current plus the shield current. The common mode current interferes with communication network, and low voltage/low current signals. Common mode current can flow through motor bearing, resulting premature motor failure.

Table 2.1 (Continued)

Item	Issues	Causes	Discussion
2	Premature VFD cable failure		VFD cable will have the following additional features: Overall shield, Symmetrically designed conductors, and better insulation. Better insulation is necessary to increase resistance to high voltage peaks and high dv/dt.
3	Malfunction of other vital shipboard equipment including shutdown		
4	Malfunction of VFD related equipment		
5	Proliferation of harmonic in the LV distribution and MV distribution and effecting auxiliary systems and contributing to malfunction		

2.2 Shipboard Electrical Propulsion System With Adjustable Speed Drive and Ship Service Power With Transformer and Motor-Generator (MG) Set (Figure 2.1 and 2.2)

The propulsion power requirement is usually much higher than the other ship service loads. The propulsion load with variable frequency drives and transformer, the current inrush can create a disturbance in the main bus. The shipboard power system must be designed so that any combination of generators can support the load requirements without any concern of high current inrush.

If a large transformer is used, pre-magnetizing system should be considered current inrush. The pre-magnetization system initially energizes the secondary of the propulsion transformers prior to connecting the transformer primary to the main bus. This in effect pre-magnetizes the transformer core and aligns the transformer magnetic field, so that the starting current inrush is negligible when the transformer main circuit breaker is closed.

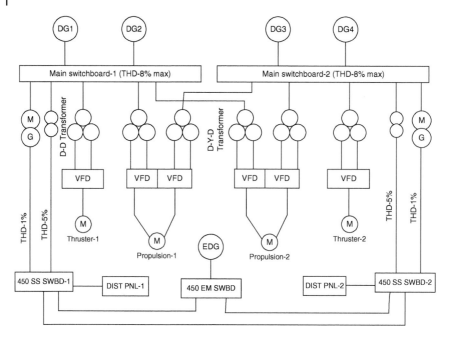

Figure 2.1 Typical electrical propulsion with variable speed controller and ship service switchboard is fed from a transformer and MG set (MG set is for clean power consideration).

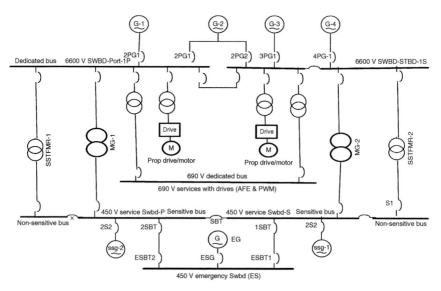

Figure 2.2 Typical Variable Speed Electric Propulsion System With Ship Service Transformer For Non-Sensitive Loads And MG Set For Sensitive Loads.

There are four ship service generators, two ship service switchboards, one emergency generator, and one emergency switchboard. Each ship service switchboard is connected to two ship service generators. This set-up is for ships with a large service load demand, such as cruise ships. Due to the number of generators, the load distribution and management of four ship service generators and one emergency generator can be complex. However, the design has proven to be very successful. The parallel operation and necessary protective control and their interlocking must however be clearly understood by the operator as shown:

- Ship service bus tie breaker section
- Emergency switchboard—120 V section
  - Emergency lighting
  - Other 115 V emergency loads
  - Other 115 V services
- Emergency switchboard—480 V section
  - Emergency generator circuit breaker
  - 480 V emergency loads
  - Ship service bus tie breaker section

Figures 2.1 and 2.2 are typical shipboard power generation and distribution systems delivering power to a lube oil pump. The main generator is delivering 6600 V power to a 6600 V switchboard. The 6600 V switchboard is delivering power to the 480 V ship service switchboard through a step down transformer. Motor generator is also added as an alternative to ensure the 480 V ship service bus type-1 power quality.

2.3 VFD System Design Verification

A design is to meets fundamentals of the requirements by contract specification, which is presentation in an electrical one-line diagram such as Figure 2.1. The basic verification is outlined in the USCG regulations called failure mode and effect analysis (FMEA) supported by qualitative failure analysis (QFA) and design verification test procedure (DVTP).

Four 6600 V generators are to generate power by using any one of the three generators keeping one generator spare to meet regulations.

The 6600 V switchboard is shown with bus-tie circuit breaker. This is a requirement. There are ways to meet this requirements which will be further discussed in various sub-sections in this chapter.

- 6600 V 3 phase, 60 Hz switchboards
- 450 V Ship service switchboard. The power is delivered by transformer 6600 V/450 V, three-phase transformers

- 450 V Vital power distribution panel with circuit breaker for the auxiliary motor. The motor feeder must be of sufficient size to ensure load carrying capability. The motor controller overload thermal unit must be sized to protect the system from overload. The circuit breaker in the 450 V distribution panel must be sized to protect the motor from normal overload as well as the overload from the short circuit condition (See Figure 2.2).

The 6600 V generation and 450 V distribution system shown with item numbers and design step numbers in reverse order to deal with each design step. This is to be designed and developed with full understanding of the design process.

In this example, there are four ship service generators, two ship service switchboards, one emergency generator, and one emergency switchboard. Each ship service switchboard is connected to two ship service generators. This set-up is for ships with a large service load demand, such as cruise ships. Due to the number of generators, the load distribution and management of four ship service generators and one emergency generator can be complex. However, the design has proven to be very successful. The parallel operation and necessary protective control and their interlocking must however be clearly understood by the operator.

Port—Ship service switchboard—480 V section

- Generator breakers
- Motor controllers
- Ship service load section for machinery auxiliaries, HVAC, deck machinery, and electrical services
- Shore power (open option of feeding from ship service or emergency switchboard)
- Emergency bus tie

Starboard—Ship service switchboard—480 V section

- Generator breakers
- Motor controllers
- Ship service load section for propulsion auxiliaries, HVAC, deck machinery, and electrical services
- Shore power (open option of feeding from ship service or emergency switchboard)
- Emergency bus tie

In this example, there are four ship service generators, two ship service switchboards, one emergency generator, and one emergency switchboard. Each ship service switchboard is connected to two ship service generators. This set-up is for ships with large service load demand, such as cruise ships. Due to the number of generators, the load distribution and management of four ship service generators and one emergency generator can be complex.

However, the design has proven to be very successful. The parallel operation and necessary protective control and their interlocking must however be clearly understood by the operator.

Port—Ship service switchboard—480 V section

- Generator breakers
- Motor controllers
- Ship service load section for machinery auxiliaries, HVAC, deck machinery, and electrical services
- Shore power (open option of feeding from ship service or emergency switchboard)
- Emergency bus tie

Starboard—Ship service switchboard—480 V section

- Generator breakers
- Motor controllers
- Ship service load section for propulsion auxiliaries, HVAC, deck machinery, and electrical services
- Shore power (open option of feeding from ship service or emergency switchboard)
- Emergency bus tie

Starboard—Ship service switchboard—120 V section

- 115 V lighting
- Other 115 V services

2.4 Typical DP ASD Propulsion, and Thrusters All With VFD Drives (Figure 2.3)

In this example, there are four ship service generators, two ship service switchboards, one emergency generator, and one emergency switchboard. Each ship service switchboard is connected to two ship service generators. This set-up is for ships with large service load demand, such as cruise ships. Due to the number of generators, the load distribution and management of four ship service generators and one emergency generator can be complex. However, the design has proven to be very successful. The parallel operation and necessary protective control and their interlocking must however be clearly understood by the operator.

There are four ship service generators, two ship service switchboards, one emergency generator, and one emergency switchboard. Each ship service switchboard is connected to two ship service generators. This set-up is for ships with a large service load demand, such as cruise ships. Due to the number

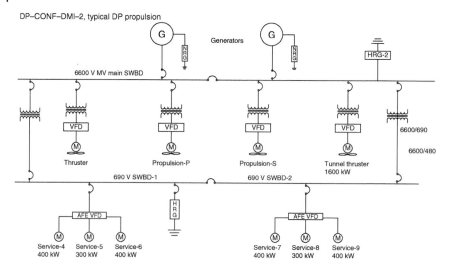

DP–CONF–DMI–2, typical DP propulsion

Figure 2.3 Typical DP ASD Propulsion, Thrusters, and VFD Drive Services.

of generators, the load distribution and management of four ship service generators and one emergency generator can be complex. However, the design has proven to be very successful. The parallel operation and necessary protective control and their interlocking must however be clearly understood by the operator.

Port—Ship service switchboard—480 V section

- Generator breakers
- Motor controllers
- Ship service load section for machinery auxiliaries, HVAC, deck machinery, and electrical services
- Shore power (open option of feeding from ship service or emergency switchboard)
- Emergency bus tie

Starboard—Ship service switchboard—480 V section

- Generator breakers
- Motor controllers
- Ship service load section for propulsion auxiliaries, HVAC, deck machinery, and electrical services
- Shore power (open option of feeding from ship service or emergency switchboard)
- Emergency bus tie

3

VFD Motor Controller for Ship Service Auxiliaries

The use of variable frequency drive (VFD) motor controllers are gaining popularity in place of across the line starters, soft starters, and reduced voltage starters due to operational advantages. The VFDs are being used on small vessels auxiliary loads such as offshore support vessels. The use of VFD in a small vessel with low power generation and weak distribution bus often creates electrical noise issues such as harmonics, affecting the entire electrical system.

The use of VFD is recommended; however, considerations should be made such as:

a) Do not use VFD in small vessels with weak power generating bus.
b) Do appropriate analysis, and calculations to establish that the VFD-related THD will be below 5% under all operating conditions.
c) Establish that the use of VFD will have no adverse effects on the electrical system.
d) The VFD-related internal components should not be grounded at any point. However, if component grounding is necessary, appropriate measures must be taken, so that the circulating current does not create an adverse effect.
e) Do not use neutral grounded low voltage system (grounded WYE).
f) If neutral grounded low voltage system 120 V, three-phase, four-wire system is used, the load distribution must be properly balanced so that the neutral current circulation is within allowable limits.
g) The system must be monitored and alarmed if the neutral current circulation is unacceptable.
h) Use proper VFD cable.
i) Use proper VFD cable segregation.
j) Identify all VFD cables with proper circuit designation.
k) Use approved VFD cable termination procedure.

VFD Challenges for Shipboard Electrical Power System Design, First Edition.
Mohammed M. Islam.
© 2020 by The Institute of Electrical and Electronics Engineers, Inc.
Published 2020 by John Wiley & Sons, Inc.

3.1   480 V System with VFD Auxiliaries (Low Power-Weak Distribution Bus)

The Figure 3.1 is used to illustrate the VFD applications in an electrical system where the power generation is limited to a few megawatts, which may be classified as a weak bus system. In this configuration, the VFD application can be challenging due to the fact that the VFD-related electrical noise phenomenon is substantial. If this type of system and the VFD application is necessary then the following guideline must be followed:

a) There can be no compromise on the ship service power stability requirements as required by the regulations.
b) Harmonic management system must be implemented at the preliminary design phase to establish that the THD-related issues are properly managed.
c) There should be no ground current circulation in the system.
d) The THD level must be measured at the delivery of the vessel to establish a base line of THD.
e) The THD measurement must be repeated annually to compare with the base line THD.
f) Integrated harmonic management system in support of the effect of individual equipment harmonics and all VFD application harmonics. See Figure 3.1.

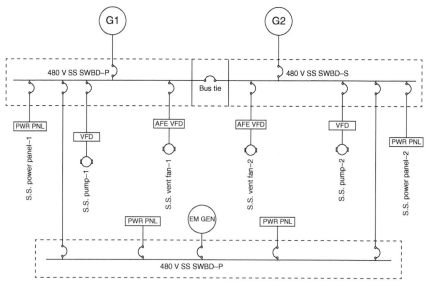

Low voltage 480 V and low power generation with VFD applications

Figure 3.1 Two 480 V Generators With Low Voltage VFD Auxiliary System (Low Power-Weak Distribution Bus) – Less Than 5% THD At The 480 V Distribution Bus.

3.2 6600 V Generation and 690 V Distribution with VFD on Both Buses

This configuration is with VFD on 6600 V bus and VFD on 690 V bus.

The 480 V ship service distribution bus is not shown as this set-up is to explain the harmonic effects. It is also assumed that the 480 V has no VFD installed for ship service auxiliaries.

If this type of system and the VFD application are necessary, then the following guideline must be followed:

a) The 6600 V can be considered as dedicated bus, which has no direct influence on the 480 V ship service bus. For this configuration, the THD is allowed to be above 5%, up to 8%.
b) The 690 V can be considered as dedicated bus, which has no direct influence on the 480 V ship service bus. For this configuration, the THD is allowed to be above 5%, up to 8%.
c) Harmonic management system must be implemented at the preliminary design phase to establish that the THD-related issues are properly managed.
d) There should be no ground current circulation in the system.
e) The THD level must be measured at the delivery of the vessel to establish a base line of THD.
f) The THD measurement must be repeated annually to compare with the base line THD.
g) Integrated harmonic management system in support of the effect of individual equipment harmonics and all VFD application harmonics. See Figure 3.2.

3.3 450 V Ship Service Bus With 6 Pulse VFD and Active Front End (AFE) VFD

The Figure 3.3 is an acceptable arrangement for the use of VFD in the 450 V ship service distribution system for the following reasons:

a) There is a galvanic isolation in the system due to the use of transformer of converting 6600 V to 450 V system.
b) As long as the bus system is considered a dedicated bus. The dedicated bus as shown in the figure simply has no ship service auxiliary loads.
c) Harmonic management can be accomplished as the distribution system has no other non-VFD load.

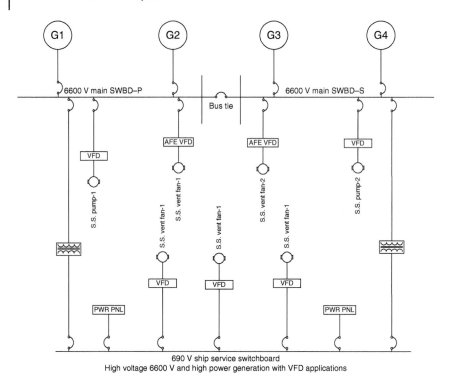

690 V ship service switchboard
High voltage 6600 V and high power generation with VFD applications

Figure 3.2 Four Generators 6600 V Generation and 690 V distribution with VFD on both buses.

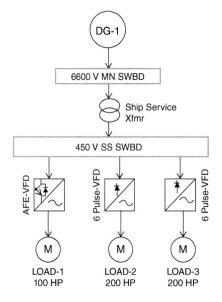

Figure 3.3 Typical 450 V Distribution bus With delta–wye–delta transformer with common rectifier for multiple inverters and braking resistors.

3.4 A 690 V Distribution System With Delta–Wye–Delta Transformer Isolation

This configuration comes with the following in support of the harmonic management:

a) Delta–wye–delta transformer configuration brings harmonic cancellation by phase shifting.
b) One rectifier is used for multiple loads by establishing a common DC bus.
c) One dynamic braking resistor system to support multiple loads. See Figure 3.4.

Figure 3.4 Typical VFD feed with delta–wye–delta transformer and dynamic braking resistors.

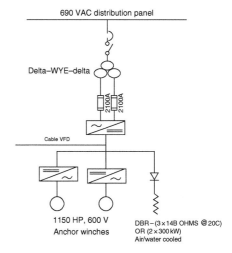

690 VAC distribution panel

Delta–WYE–delta

2100A 2100A

Cable VFD

1150 HP, 600 V
Anchor winches

DBR – (3 × 14B OHMS @ 20C)
OR (2 × 300 kW)
Air/water cooled

3.5 A 690 V Ship Service Bus Powering VFD With Details of VFD Feeder Cable and Motor Feeder Cable

It is very important to identify and specify proper VFD cable, which is the cable from the VFD to the motor. The VFD cable termination requirements should be as recommended by the VFD manufacturer. Refer to the Figure 3.5 for additional details.

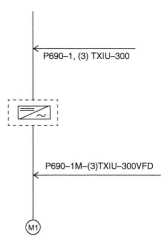

P690–1, (3) TXIU–300

P690–1M–(3)TXIU–300VFD

(M1)

Winch, 1000 HP, 600 V
690 V distribution

Figure 3.5 Sample 690 V VFD Auxiliary System Feeder showing VFD cable.

4

Shipboard Power System With LVDC and MVDC for AC and DC Application

The direct current LVDC (up to 1200 VDC) and MVDC (8000 VDC) are being considered for certain applications. The LVDC and MVDC distribution may not be cost effective, as the technology is not matured enough for commercial applications. However, there are major advantages of using the direct current main distribution bus which may be suitable for some applications, including the VFD applications.

Refer to the following Navy standards for recommended use:

Refer to MIL-STD-1399 (Navy) for LVDC (Up to 1000 VDC)

Refer to MIL-STD-1399 (Navy for MVDC (6000 VDC–18000 VDC)

4.1 The Advantage of Using DC Distribution

The advantages of a direct current (LVDC & MVDC) main distribution system include:

a) The prime mover speed is decoupled from the power quality of the bus. The generator can be optimized for each type of prime mover without having to incorporate reduction gears or speed increasing gears; generators are not restricted to a given number of poles. The speed can even vary across the power operating range of the prime mover.

b) Power conversion equipment can operate at higher frequencies resulting in smaller transformers and other electromagnetic devices.

c) Unlike AC cables, the full cross section of a DC conductor is effective in the transmission of power. Furthermore, power factor does not apply to DC systems and therefore does not influence cable size. Depending on the selected bus voltage, cable weights may decrease for a given power level.

VFD Challenges for Shipboard Electrical Power System Design, First Edition.
Mohammed M. Islam.
© 2020 by The Institute of Electrical and Electronics Engineers, Inc.
Published 2020 by John Wiley & Sons, Inc.

d) Power electronics can control fault currents to levels considerably lower than with AC systems employing conventional circuit-breakers. Lower fault currents reduce damage during faults.

e) The paralleling of power sources only requires voltage matching and does not require time-critical phase matching. This enables generator sets to come on line faster after starting, thereby reducing the aggregate amount of energy storage needed to provide power while another generator set is brought online.

f) High speed prime mover coupled to high speed generators that produce higher than 60 Hz frequency power are easily accommodated.

4.2 Typical Direct Current Power Distribution System (Refer to Figure 4.1)

The high speed engine is directly connected to the generator and the AC/DC convereter. The DC power is supplied to the DC distribution switchboard.

The DC distribution switchboard supplies power to various ship service loads with appropriate DC to AC converters such as:

- DC/AC to 480 VAC load center
- DC/AC VFD for multiple AC motors
- DC/AC VFD for propulsion power

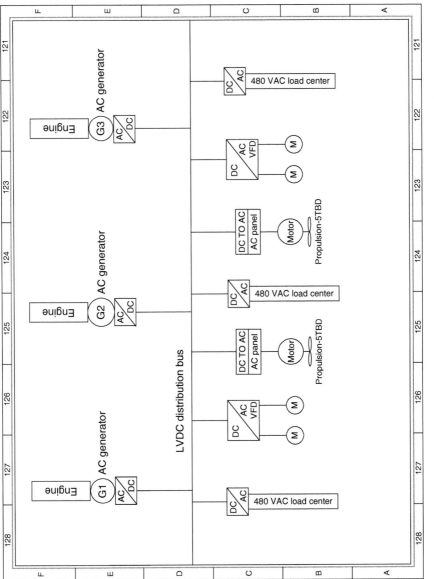

Figure 4.1 Typical direct current main distribution electrical one line diagram for commercial ship applications.

5

Shipboard VFD Application and System Grounding

The shipboard power system is different from the shore-based power distribution with neutral line and ground line. The shipboard power system is ungrounded. The shore-based power distribution is with neutral line and ground line. However, there are special cases on ships where the grounded distribution is used in a controlled manner, such as shipboard workshop equipment.

In some installations the shipboard power system, particularly the low voltage grounded distribution system is used along with the 450 V ungrounded system. This feature of ungrounded ship service power system distribution and low voltage grounded distribution system is presented in view of the VFD installation (See Figure 5.1).

5.1 VFD Motor Controller and Grounding

Shipboard VFD motor controller installation in an ungrounded power system EMC/EMI noise played a vital role. To mitigate VFD-generated EMI noise and to prevent high frequency current affecting other equipment, the understanding of equipment grounding, drain wire requirements, and cable shielding, as well as termination of the main conductors, drain wires, and shielding is very important.

The following termination guidelines are recommended:

a) VFD drain wire (ground wire) should have a positive ground connection at the VFD enclosure as well as the motor enclosure, ensuring low resistance path.
b) The ground wire should of the same size as the power cable to maintain the same current level protection.

VFD Challenges for Shipboard Electrical Power System Design, First Edition.
Mohammed M. Islam.
© 2020 by The Institute of Electrical and Electronics Engineers, Inc.
Published 2020 by John Wiley & Sons, Inc.

1. Typical residential electrical service with phase to phase 240 V and phase to neutral 120 V service
2. G is a bare conductor G is not a current carrying conductor.
3. In the main panel G and N are jumpered.
4. In the load enclosure the ground conductor is connected to the enclosure as a safety ground.. the load is connected to L1 and N.. if one line is broken and comes in contact with the enclosure, it will establish a current path through the G connection. Therefore the G is considered current carrying path.

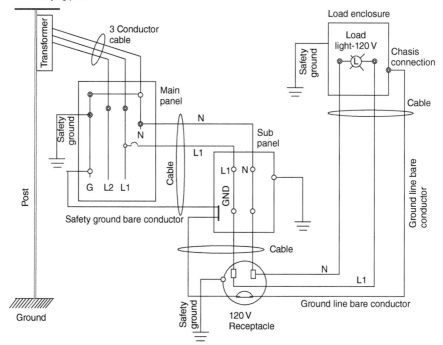

Figure 5.1 Typical Land Based Power System with Neutral And Ground Configuration.

c) The cable shielding should be terminated in the VFD enclosure, ensuring 360 degree positive connection. Cable shield provides a positive return path for common mode noise.

d) The cable shielding should be terminated at the motor enclosure ensuring 360 degree positive connection.

The shipboard ungrounded electrical distribution system can develop capacitive ground through system level capacitance.

There are cases where the shipboard power generation and distribution have a three-phase and four-wire system, the fourth wire being the system ground connected to the hull. This four-wire system will be addressed on a limited basis in this book.

In general, the shipboard ungrounded electrical system provides for the following:

a) To prevent contact with a dangerous voltage if electrical insulation fails or electric power system fault occurs. To reduce the arc or flash hazard to personnel who may have accidently caused this, or who happen to be in close proximity to the ground fault.
b) Efforts to minimize equipment damage from fault overvoltage, such as switchgear, drives, transformers, cable, rotating machines.
c) Reduction of static electricity build-up on rotating machines.
d) Reduction of common mode transient overvoltage.
e) Eliminate arcing ground-faults and transient overvoltage.
f) Provide an equal voltage reference to all phase conductors to balance phase voltages.
g) Reduce harmonics at all frequencies with respect to ground for a clean reliable control operation.
h) Provide energy absorption and protection for equipment within the CBEMA curve.
i) Provide an early warning signal and ground indication for arc-fault hazard warning.
j) Remain operational during an event for single-phase fault.

For shipboard low voltage ungrounded power generation and distribution system ground detection requirements refer to IEEE-45. For medium voltage distribution ground detection system with resistance grounding requirements refer to IEEE-142.

For insulation monitoring system along with ground detection system refer to IEC-61577 .

The ground detection systems are outlined here for the design engineers to better understand the limits of each application. The ground detection system is very much a part of the safety considerations, therefore, any ground detection system which is selected must be properly analyzed and understood for the protection of the shipboard system, both its equipment as well the operators.

5.2 Grounding for the VFD Motor Controller EMI

In general, the VFD motor controllers generated noise management filters are mostly supported by ground reference. For a shipboard ungrounded system these filters may create additional system integrity problems, which must be properly engineered for such a filter application.

The VFD motor controller cable is called the VFD cable. The VFD cable consists of the main three-phase power conductors, three symmetric drain wires, and power cable shielding. The drain wires must be connected to a specific ground path. The cable shielding must be connected to a specific ground path as recommended by the VFD supplier. See Figures 5.2 and 5.3.

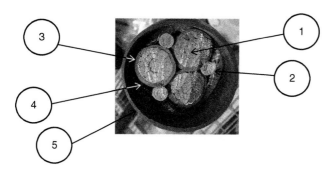

1. Three power conductors-flexible stranded tinned copper per IEEE-45/1580
2. Three ground conductors-green color soft annealed flexible stranded copper per IEEE-45/1580
3. Insulation 2 kV class. Color gray
4. Overall shield – tinned copper braid with aluminium/polyster tape
5. Jacket-flame retardant, oil, abresion, chemicaland sunlight resitant thermosetting compound. Color-black (can be of other color to differentiate from other power cable)

Figure 5.2 Typical VFD Power Cable Shield Grounding.

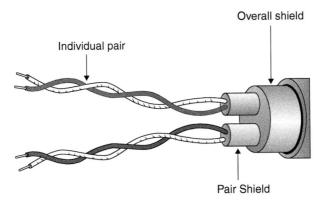

Figure 5.3 Typical VFD Control And Signal Cable Shield Grounding.

The grounding conductor associated with power grounding, control grounding, or signal grounding has the properties of resistance, capacitance, and inductance.

- The conductor shield and the drain-wire resistance is a function of the material, its length and its cross-sectional area.
- The capacitance is determined by its geometric shape, its proximity to other conductors and dielectric.
- The inductance is a function of size, geometry, length, and permeability of the metal.
- The impedance of the grounding system is a function of resistance, inductance, capacitance and frequency.

For a typical length of grounding line, the wire strap (bond strap) has an impedance which is much less than circular wire. The impedance of both bond strap and circular wires increase with an increase in frequency and become significant in higher frequency. At some frequency the inductive reactance $j\omega L$ will be equal to capacitive reactance $1/j\omega c$ and the circuit becomes resonant. Therefore, the resonant frequency is

$$f = 1/2\pi\sqrt{LC}$$

5.3 NVIC 2-89 Requirement for Ground Detection for Ungrounded System and Grounded System (Extract Only)

Ground Detection General

Grounds can be a source of fire and electric shock. In an ungrounded system, a single ground has no appreciable effect on current flow. However, if low resistance grounds occur on conductors of different potentials, very large currents can result. In grounded system. a single low impedance ground can result in large fault currents, To provide for the detection of grounds, the regulations require that ground detection, means be provided for each electric propulsion system, each ship's service power system, each lighting system, and each power or lighting, system that is isolated from the ship's service power and lighting system by transformers, motor generator sets, or other device. This indication need not be part of the main switchboard but should be co-located with the switchboard (i.e. at the engineering control console adjacent to the main switchboard).

Ground Detection Ungrounded System

The indication maybe accomplished by a single bank of lights with a switch which selects the power system to be tested, or by a set of ground detector lights for each system monitored. In an ungrounded three-phase system, ground detection lamps are used. The ground lamps are connected in a "wye" configuration with the common point grounded. A normally-closed switch is provided in the ground connection.

If no ground is present on the system, each lamp will see one-half of the phase-to-phase voltage and will be illuminated at equal intensity If line "A" is grounded by a low impedance ground, the lamp connected to line "A" will be shunted out and the lamp will be dark. The other two lamps will be energized at phase-to-phase voltage and will be brighter than usual. If a low resistance ground occurs on any line, the lamp connected to that line will be dimmed slightly and the other two lamps will brighten slightly. The switch is provided to aid in detecting high impedance grounds that produce only a slight voltage shift. When the ground connection is opened by the switch, the voltage across each lamp returns to normal [phase voltage) and each lamp will have the same intensity. This provides a means to observe contrast between normal voltage and voltages that have shifted slightly. Lamp wattages of between 5 W and 25 W when operating at one-half phase-to-phase voltage (without a ground present) have been found to perform adequately giving a viewer adequate illumination contrast for high impedance grounds. Should a solid ground occur, the lamps will still be within their rating and will not be damaged. For lesser grounds, the lumen output of the lamps will vary approximately proportional to the cube of the voltage. This exponential change in lamp brightness (increasing in two and decreasing in one) provides the necessary contrast.

Ground Detection—Grounded System

On grounded dual voltage systems, an ammeter is used for ground detection. This ammeter is connected in series with the connection between the neutral and the vessel ground. To provide for the detection of high impedance grounds with correspondingly low ground currents; the regulations specify an ammeter scale of 0 to 10 A However the meter must be able to withstand; without damage, much higher ground currents, typically around 500 A. This feature is usually provided by the use of a special transducer such as a saturable reactor in the meter circuit. Some ammeters use a non-linear scale to provide for ease in detecting movement at low current values.

Other types of solid-state devices are becoming available that can provide ground detection. They should not be prohibited, but should be evaluated to determine that they are functionally equivalent to the lights and ammeters historically used. Some systems also include a visual and or audible alarm at a preset level of ground current.

5.4 System Grounding – IEEE-45 and USCG Requirements

5.4.1 Shipboard Power System Grounding IEEE-45 Recommendations

IEEE-45-2002: Ground detection for each ungrounded system should have a monitoring and display system that has a lamp for each phase that is connected between the phase and the ground. This lamp should operate at more than 5 W and less than 24 W when at one-half voltage in the absence of a ground. The monitoring and display system should also have a normally closed, spring return-to-normal switch between the lamps and the ground connection. If lamps and continuous ground monitoring utilizing superimposed DC voltage are installed, the test switch should give priority to continuous ground monitoring and should be utilized only to determine which phase has the ground fault by switching the lamp in. With continuous ground monitoring tied into the alarm and monitoring system, consideration will be given to alternate individual indication of phase to ground fault. If lamps with low impedance are utilized, the continuous ground monitoring is reading ground fault equivalent to the impedance of the lamps, which are directly connected to the ground.

Where continuous ground monitoring systems are utilized on systems where nonlinear loads (e.g., adjustable speed drives) are present, the ground monitoring system must be able to function properly.

IEEE-45-2002 CLAUSE-5.9.7.2 Ground Detection Lamps on Ungrounded Systems

Ground detection for each ungrounded system should have a monitoring and display system that has a lamp for each phase that is connected between the phase and the ground. This lamp should operate at more than 5 W and less than 24 W when at one-half voltage in the absence of a ground. The monitoring and display system should also have a normally closed, spring return-to-normal switch between the lamps and the ground connection.

If lamps and continuous ground monitoring utilizing superimposed DC voltage are installed, the test switch should give priority to continuous ground monitoring and should be utilized only to determine which phase has the ground fault by switching the lamps in.

With continuous ground monitoring tied into the alarm and monitoring system, consideration will be given to alternative individual indication of phase to ground fault. If lamps with low impedance are utilized, the continuous ground monitoring is reading ground fault equivalent to the impedance of the lamps, which are directly connected to the ground.

Where continuous ground monitoring systems are utilized on systems where nonlinear loads (e.g., adjustable speed drives) are present, the ground monitoring system must be able to function properly.

5.4.2   Typical Ground Detection System of Ungrounded Distribution - USCG Requirements

See Figure 5.4.

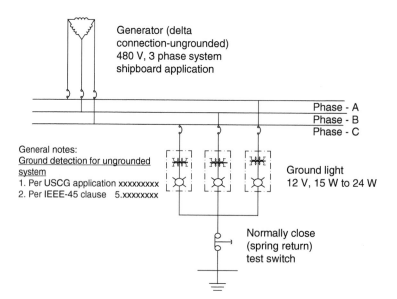

Generator (delta connection-ungrounded) 480 V, 3 phase system shipboard application

Phase - A
Phase - B
Phase - C

General notes:
Ground detection for ungrounded system
1. Per USCG application xxxxxxxxx
2. Per IEEE-45 clause 5.xxxxxxxxx

Ground light
12 V, 15 W to 24 W

Normally close (spring return) test switch

Figure 5.4 Typical Shipboard Ground Detection System For Ungrounded Generation And Distribution (Type-1).

5.5 Practical Consideration of Selecting Specific Type of High Resistance Grounding (HRG) For Shipboard Ungrounded Power System

Shipboard ungrounded power system is used, because under single-phase short the system is considered operational for a short period. The single-phase fault is limited by the capacitance of the healthy phases. However, it must be noted that during the single-phase ground fault the voltage to ground of the ungrounded phases increase by the factor of square root of 3. For a typical 6600 V system the line to ground (LG) voltage can be = 3810 V. This overvoltage may lead to insulation stress for all electrical equipment within that system. If the insulation rating is not sufficient to withstand the overstress voltage, the equipment will get overheated and ultimately can fail. If a second ground fault occurs before clearing the the first fault, this can lead to an arcing fault. Therefore, for a shipboard ground detection system, the single-phase fault and phase to phase fault must be detected as soon as possible and protective measures must be taken to avoid arcing explosion and equipment malfunction onboard ship.

The resistance value should be selected slightly higher than the capacitive charging current of the system to ensure the lowest level of fault current flows at which the system over-voltage is limited. The high resistance grounding system is developed by creating a ground point by using transformer with WYE primary. There are many different types of setting the ground point, such as direct ground connection or ground connection through grounding resistors.

The zigzag transformer is also used where the grounding resistror is connected between the neutral of the primary winding and ground. The grounding resistor must be rated for 3180 Vm, which will make the resistor large and expensive. The grounding resistor must be rated for 3180 Vm, which will make the resistor large and expensive.

5.6 IEEE-142 Ground Monitoring Requirements

Extract from IEEE-142-2007, Recommended Practice for Ground Detection System

Clause 1.4.3.1 High-resistance Grounding

High-resistance grounding employs a neutral resistor of high ohmic value. The value of the resistor is selected to limit the current Ir to a magnitude equal to or slightly greater than the total capacitance charging current, 3 Ico.

Typically, the ground-fault current is limited to 10 A or less, although some specialized systems at voltages in the 15 kV class may require higher ground-fault levels.

In general, the use of high-resistance grounding on systems where the line-to-ground fault exceeds 10 A should be avoided because of the potential damage caused by an arcing current larger than 10 A in a confined space (see Foster, Brown, and Pryor).

Several references are available that give typical system charging currents for major items in the electrical system (see *Electrical Transmission and Distribution Reference Book*; Baker). These will allow the value of the neutral resistor to be estimated in the project design stage. The actual system charging current may be measured prior to connection of the high-resistance grounding equipment following the manufacturer's recommended procedures.

5.7 IEC Requirements

System Insulation Monitoring:

IEC requirement of ground detection system included system level insulation monitoring system which goes beyond the requirements for ground detection system. The details of IEC-61577-8 is below:

IEC-61577-8 (2007)

SECTION-4 REQUIREMENTS

THE FOLLOWING REQUIREMENTS AS WELL AS THE REQUIREMENTS OF 61577-1 SHALL APPLY

- 4.1 Insulation monitoring devices shall be capable of monitoring the insulation monitoring of IT systems including symmetrical and asymmetrical components and to give a warning if the insulation resistance between the system and earth falls below predetermined level.
  Note-1: The symmetrical insulation deterioration occurs when the insulation resistance of all conductors in the system to be monitored decreases (approximately similarly). An asymmetric insulation deterioration occurs when the insulation resistance of (one conductor for example) occurs substantially more than the other conductor.
  Note-2: So called earth fault relays using a voltage asymmetry (voltage shift) in the presence of an earth fault as the only measurement criteria, are not insulation monitoring devices in the interpretation of IEC 61577.
  Note-3: A combination of several measurement methods including asymmetric monitoring may become necessary for fulfilling the task of monitoring special conditions on the system.
- 4.2 The insulation monitoring system shall comprise a test device, or be provided with a means for connection of a test facility for detecting whether

the insulation monitoring device is capable of fulfilling its functions. The system to be monitored shall not be directly earthed and the devices shall not be suffered damage. This test is not intended for checking the response value.

- 4.3 Contrary to IEC-61577-1, the PE connection of insulation monitoring device is a measuring connection and may be treated as functional earth connection (FE). If the IMD has additional parts which are grounded for protection functions this connection shall be

5.8 Recommendations for System Grounding

The grounding system should be designed to minimize the magnitude of ground fault current flowing in the hull structure. Methods of grounding power distribution systems should be determined considering the following:

a) Grounded systems reduce the potential for transient overvoltage.

b) To the maximum extent possible, system design should allow for continuity of service under single- line-to-ground fault conditions, particularly in distribution systems supplying critical ship's service loads.

c) Systems should be designed to minimize the magnitude of ground fault currents flowing in the hull structure.

d) Balance phase voltages with respect to ground for reliable operation and equipment protection

To satisfy these criteria, more than one type of grounding system will be necessary for shipboard specific installation requirements. However, it is recommended that systems be designed per one of the following grounding philosophies:

i) Ungrounded, with all current-carrying conductors completely insulated from ground throughout the system. Ungrounded systems should have provisions for continuous ground fault monitoring. Depending on the system requirement, additional features such as alarm and protection may be considered. However, for high resistance ground system is preferred over ungrounded system. As outlined in (ii):

ii) High-resistance grounded such that single-line-to-ground faults are limited to 5 Amp, maximum. High-resistance grounded systems should have provisions for continuous ground fault monitoring. In addition, cables or raceways containing power conductors should be provided with equipment grounding conductors sized in accordance with NEC Table 250-122 to minimize the possibility of ground fault currents flowing in the hull structure. Depending on the system requirement, additional features such as alarm and protection may be considered.

iii) Solidly grounded. Solidly grounded designs should be limited to systems supplying noncritical loads, such as normal lighting, galley circuits, and

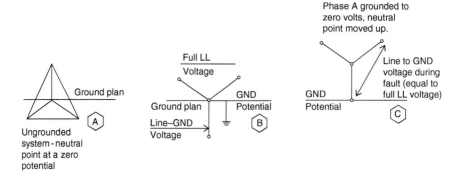

Figure 5.5 Typical Ungrounded Distribution Vector Diagram (Type-1).

so on. When a solidly grounded distribution system is used, the neutral conductor should be full-sized to preclude overheating due to harmonic distortion from nonlinear loads. (See Clause 33 for restrictions in hazardous locations.) In addition, cables or raceways containing power conductors should be provided with equipment grounding conductors sized in accordance with NEC Table 250-122 to minimize the possibility of ground fault currents flowing in the hull structure.

I) Depending on the system requirement, additional features such as alarm and protection may be considered. See Figures 5.5 and 5.6.

5.9 Insulation Monitoring

5.9.1 Insulation Monitoring Device (IMD)

The IMD is the device of choice for the protection of floating systems. IMDs come in two styles: a) passive; and b) active devices.

5.9.2 Passive Insulation Monitoring Device (IMD)

The most known passive insulation monitoring device is the three light bulb system in 480 V ungrounded systems. Three lamps are connected on their secondary side together and from there to ground (Star or Y configuration). Each lamp is then connected to the respective phases L1, L2, and L3. In a healthy system, all three lights will burn with the same intensity. In case of a ground fault, the faulted phase will assume a value close to ground potential. The respective light will dim, while the other two will brighten up. The light bulb system often does not offer additional trip indicators for remote alarms. It also needs to experience a serious fault condition before people are becoming aware that something is going wrong. Even worse, symmetrical ground faults (A balanced fault on all three-phases) will not be detected.

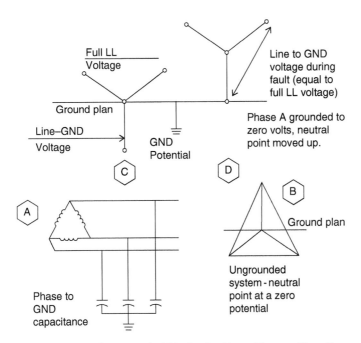

Figure 5.6 Typical Ungrounded Distribution Vector Diagram (Type-2).

5.9.3 Active Insulation Monitoring Device (IMD)

The active Insulation Monitoring Device is considered to be an online megger. It will be connected via pilot wires between the system and ground. A constant measuring signal will be sent from the IMD into the power wires. It will spread out evenly into the secondary side of the supply transformer and the attached loads. If this signal finds a break through path to ground, it will take this path of least resistance and return to the monitor. The IMDs internal circuitry will process the signal and trip a set of indicators when the set point is increased. IMDs measure in Ohms (Resistance) and not in Amps (Current). A ground fault will be indicated as "insulation breakdown".

> Healthy system is usually multiple kilo ohms or mega ohms
> Low insulation = ground fault = less than one kilo ohm or low ohm range

A power system's overall resistance depends on the number of loads, the type of insulation used, the age of the installation, the environmental conditions etc. A typical question when it comes to floating deltas is always: "Where should my set or trip level be?" The typical "ball park" figure for industrial applications is 100 Ohms per volt.

Example: A monitor in a 480 V delta system would be set to trip at 480 × 100 = 48 kilo Ohms. Please be advised that this figure cannot be used for all situations.

Example: A customer has meggered a motor and figures that his system must be at around 1 Meg Ohm insulation. The insulation monitor keeps alarming and indicates lower levels than previously assumed. The answer is simple: It was forgotten to take into consideration that there are 10 of these motors connected to the same system. The IMD will measure and indicate the OVERALL resistance of the system. Here we are dealing with 10 parallel resistances of 1 Meg Ohm each. The overall resistance will drop to less than 100,000 Ohms in this case.

5.9.4 Insulation Monitoring System for Grounded AC Systems With VFD System

The 60 cycle ground fault relay (GFRs) have limitations when the circuitry involves VFDs (Variable Frequency Drives). Tests have shown that the typical GFR cannot keep the adjusted trip point when the system frequency changes to values below 60 cycles. Even worse, a total failure can be expected at frequencies below 12 cycles. A variable frequency drive converts the incoming AC internally into DC, which will then be modulated again into a variable cycle AC leading to the load.

Some drives might be equipped with their own internal scheme to detect ground faults which will eventually trip in the high Ampere range. Early warning or personnel protection cannot be guaranteed in this case.

VFD-EMI Filter

- The VFD often incorporates built-in EMI filters. They provide a leakage path to ground and add to the overall system leakage.

The drive uses a multiple KHZ carrier frequency. The carrier frequency can cross the gap between insulation and ground and add to the inherent leakage.

Harmonic content - Transient voltage spikes

5.9.4.1 Basic Power Systems - Ungrounded (Floating)

Floating systems are derived from a power source where there is virtually no connection to ground. 480 VAC delta configured transformers are a typical supply for a floating system. Some deltas in the mining industry can be found in hoists. 480 VAC deltas are also in wide spread use to supply 1000 Amp–2000 Amp main feeder circuits in general industrial applications. Floating systems are often used in areas where a sudden shut down must not occur. Examples are Intensive care units (ICUs) in hospitals, signal circuits, and emergency back-up systems.

5.9.4.2 The Ground Fault

The magnitude of a ground fault current in an ungrounded system is very small. It depends on the system voltage, the resistance of the ground fault causing part and the system capacitances.

5.9.4.3 Insulation Monitoring Device IMD

Ungrounded systems will not produce the amount of fault current needed to trip a common GFR. The IMD is the device of choice for the protection of floating systems. IMDs come in two styles: a) Passive and b) Active devices.

5.10 System Capacitance to Ground Charging Current Calculation (Taken From IEEE-142)

See Figures 5.7 to 5.9.

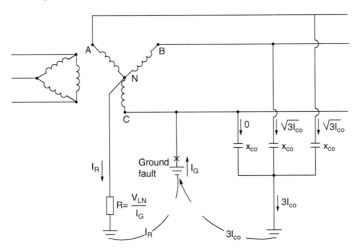

Figure 5.7 IEEE-142 (Figure 1-9 Single Line To Ground Fault On A Low Resistance Grounding System For Marine Ungrounded System).

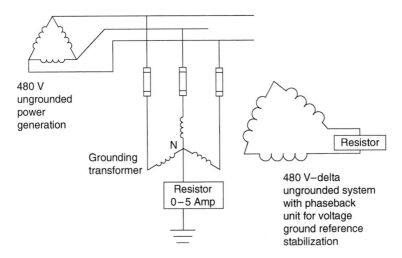

Figure 5.8 Typical Shipboard Ground Detection System.

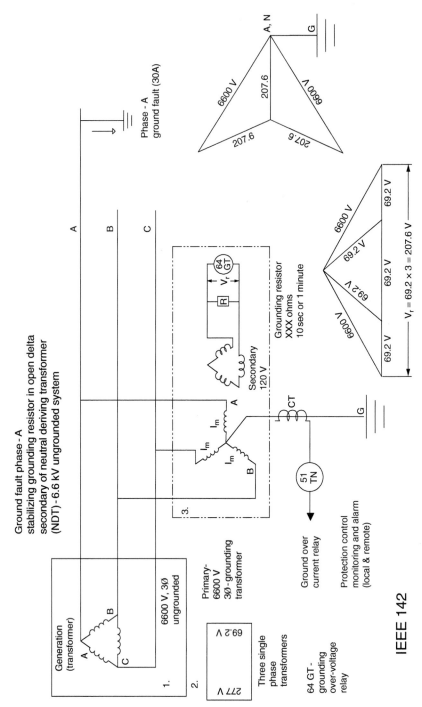

Figure 5.9 Typical Shipboard Ground Detection System.

IEEE 142

5.11   Total System Capacitance-Measurements

High Resistance Grounding is recommended for systems where power interruption resulting from single line-to-ground fault tripping is detrimental to the process

The maximum ground fault current allowed by the Neutral Grounding Resistor must exceed the total capacitance to ground charging current of the system.

The total capacitance to ground charging current of a system can be measured or estimated.

Should only be done by qualified personnel. Power should be off before making any connection before the test. The system has to be ungrounded. All components and devices used should be rated properly for the system voltage. For the test the system will be fully energized.

One line will be bolted to ground using a fast-acting fuse rated for 10 A or less, a Circuit Breaker, a Rheostat and an Ammeter all in series.

The rheostat should be set at maximum resistance and the Breaker open.

The Breaker will then be closed and the Rheostat resistance will be reduced slowly to zero. At this point the current circulating to the ground and indicated in the Ammeter is the Capacitance to Ground Charging Current of the system.

To finish the test the Rheostat will be returned to maximum resistance and the Breaker will be opened.

Please note that as with any capacitor, and here we are looking at the electrical system as a large distributed capacitor, this current will change if the physical configuration of the system is altered (i.e. by adding feeders, motors and more importantly surge arresters)

5.12   Grounding Transformer Size Calculation

5.12.1   Low Resistance Grounding-Transformer Sizing for 2400 V System

The neutral diving transformer for a shipboard ungrounded system is normally exposed to the full line-to-neutral voltage including any transients and momentary voltage excursions. Therefore, the protection of the HRG transformer should be taken under consideration.

For low resistance grounding system, at 2400 V, distribution of 300 A level, and transformer rating 2400/480 V (5:1 ratio), the secondary current will be $(100) \times (5) = 500$ A, which will circulate in the broken delta windings. Each of three grounding transformer ratings will be $(100 \text{ A}) \times (2400 \text{ V}) = 240$ kVA.

The transformer should be able to withstand maximum duration of the ground fault. Assuming ground fault duration of 10 seconds.

To limit the ground transformer in a broken delta configuration for secondary current to 500 Amps, the resistor should be rated at $(480 \times \sqrt{3}) = 831$ V, and 500 Amps, $= 831/500 = 1.66$ ohms for 10 seconds at 831 voltage rating, knowing that the broken delta voltage during a ground fault would be 831 volts. The resistor voltage rating can be at the range of 600 V–1000 V.

When the broken delta voltage during solid ground fault of 300 Amp will cause the neutral shift of $2400/\sqrt{3} = 1386$ Volts at the primary. For the ground fault of 150 Amps, the neutral shift will be half of $1386 = 693$ Volts at the primary and 415.5 at the open delta.

5.12.2 High Resistance Grounding-Transformer Sizing for 6600 V System

Based on the total capacitive charging current of 3 Amps/phase, the grounding transformer rating is $3 \times 6600 \approx 20$ kVA/phase (Note: The transformer to be rated for line to line voltage in accordance with ANSI/IEEE Std C37.101 to withstand line to line voltage underground fault conditions).

Each single-phase grounding transformer with multiple taps to select the applicable rating for the ground fault current to be slightly higher than the total capacitive charging current on the 6.6 kV system.

The transformer to be continuous rated for the 6.6 kV system to operate continuously on a single line to ground fault.

Based on the total charging current of 3 Amps/phase the single-phase grounding transformer to be selected with the nearest tap set for the transformer line current of 3 Amps (i.e for the ground current of $3 \times 3 = 9$ Amps) to provide the continuous rating of $6.6 \times 3 \approx 20$ kVA with a standard transformer ratio of 6600 V/120 V.

To ensure that the ground fault current is slightly greater than the 6.6 kV system capacitive charging current, the nearest standard transformer tap selected is for the line current of 3.3 Amps (ground current of $3 \times 3.3 \approx 10$ Amps), and this will give the transformer continuous rating of 22 kVA per phase.

5.12.3 Grounding Resistor Size Calculation

With all three single-phase transformers (each rating 22 kVA) connected in star on the primary side (6600 Volts) and neutral solidly grounded, and the secondary side connected in open delta, the secondary current is (based on the transformer primary line current of 3.3 Amps):

$$(6600 / 240) \times 3.3 = 91.5 \text{ Amps}$$

Maximum open delta voltage under ground fault is $\sqrt{3} \times 240 = 415.7$ Volts
Loading Resistor required $= 415.7/91.5 = \underline{4.5\ ohms}$
Therefore, continuous rating of the resistor is $= 415.7 \times 91.5/1000 = \underline{38\ kW}$

5.12.4   Ground Fault Protection

For the 6.6 kV system ground fault, ground fault monitoring relays are provided at the locations as detailed in the simplified system diagram shown below.

5.12.4.1   Propulsion Motor Grounding

The grounding of the propulsion motor is performed by using a resistance (R_{mas}) directly connected between the neutral of the propulsion machine and the earth because the propulsion drive is completely insulated from the HV switchboard by the propulsion transformers (galvanic isolation).

Remark : the use of a grounding transformer is not applicable to the propulsion drives due to the possible polarization of this transformer in case of DC loop ground fault.

5.12.5   The Following System Level Precaution Should be Taken Under Consideration

Propulsion motor: The presence of a Neutral Grounding Resistance (NGR) for earth fault detection makes on line monitoring of motor insulation resistance impossible (in order to do this the motor must be isolated by disconnection and the insulation resistance should be measured with a Megger). See Figure 5.10.

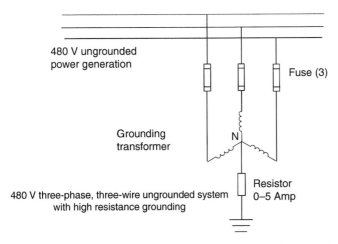

Figure 5.10 480 V System Typical High Resistance Ground Detection System.

5.13   Extract From IEEE-142-Harmonic Current Circulation

Extract from IEEE-142-2007

1.7.3.1 Circulating Harmonic Current

If generators are of identical design, there will be no significant circulation of third harmonic current while the generators are being operated at identical power and reactive current outputs. If the generators are not of identical design, there will be a third harmonic circulating current. If identical generators are operated with unequal loading, there will be a third harmonic circulating current.

Generators with two-thirds pitch windings have the minimum impedance to the flow of third harmonic currents generated elsewhere due to their low zero-sequence impedance. High-resistance grounding of the generators will adequately limit these harmonic currents. Thus, it is attractive to use high-resistance grounding on the generators, as shown in Figure 1-29, even if there are load feeders directly connected to the generator bus, and to use low-resistance bus grounding to provide selective relaying on the load feeders. Low resistance grounding of the generators at values not exceeding 25% of generator rating will normally suppress third harmonic current to adequate values even with dissimilar generators, but the variable ground-fault current available with multiple generators may pose a relay-coordination problem.

5.14   Grounding Resistor Selection Guideline per IEEE-32 Standard

---

IEEE-32 is the standard used for rating and testing neutral grounding resistors. The most important parameters to consider:

- allowable temperature rises of the element for different "on" times;
- applied potential tests; the dielectric tests, and
- resistance tolerance tests that are required.
- Time Rating- Neutral Grounding Resistor (NGR)
  IEEE-32 specifies standing time ratings for Neutral Grounding Resistors (NGRs) with permissible temperature rises above 30 °C ambient. Time ratings indicate the time the grounding resistor can operate under fault conditions without exceeding the temperature rises.

- 10-Second Rating
  This rating is applied on NGRs that are used with a protective relay to prevent damage to both the NGR and the protected equipment. The relay must clear the fault within 10 seconds.
- One-Minute Rating
  One NGR is often used to limit ground current on several outgoing feeders. This reduces equipment damage, limits voltage rise and improves voltage regulation. Since simultaneous grounds could occur in rapid succession on different feeders, a 10-second rating is not satisfactory. The one-minute rating is applied.

5.15 Grounding Resistor Duty Rating

Short time rating: Short time ratings are 0–10 second or 0–60 seconds. Since short time rated resistors can only withstand rated current for short period of time, they are usually used with fault clearing relays. The short time temperature rise for the resistive element is 760 °C.

Extended time rating: A time ratings greater than 10 minutes which permits temperature rise of resistive elements to become constant, but limited to an average not more than 90 days per year. The extended temperature rise for the resistive element is 610 °C.

Continuous rating: Capable of withstanding rated current for an indefinite period of time. The continuous temperature rise for the resistive element is 385 °C. See Figure 5.11.

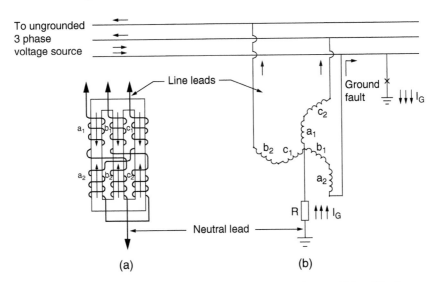

Figure 5.11 IEEE-142 (Figure 1-14(1) ZIGZAG Grounding Transformer: (a) Core Winding (b) System Connection FOR Marine Ungrounded System).

5.16 (IEEE-142 Section 1.5.2)-ZIGZAG Grounding Transformers

One type of grounding transformer commonly used is a three-phase zigzag transformer with no secondary winding. The internal connection of the transformer is illustrated in Figure 1-14(1). The impedance of the transformer to balanced three-phase voltages is high so that when there is no fault on the system, only a small magnetizing current flows in the transformer winding. The transformer impedance to zero-sequence voltages, however, is low so that it allows high ground-fault currents to flow. The transformer divides the ground-fault current into three equal components; these currents are in phase with each other and flow in the three windings of the grounding transformer. The method of winding is seen from Figure 1-14(1) to be such that when these three equal currents flow, the current in one section of the winding of each leg of the core is in a direction opposite to that in the other section of the winding on that leg. This tends to force the ground-fault current to have equal division in the three lines and accounts for the low impedance of the transformer-to-ground currents.

A zigzag transformer may be used for effective grounding, or an impedance can be inserted between the derived neutral of the zigzag transformer and ground to obtain the desired method of grounding. This transformer is seldom employed for medium-voltage, high-resistance grounding. An example of low-resistance grounding is shown in Figure 1-14(2). The overcurrent relay, 51G, is used to sense neutral current that only flows during a line-to-ground fault.

See Figure 5.12 to 5.15.

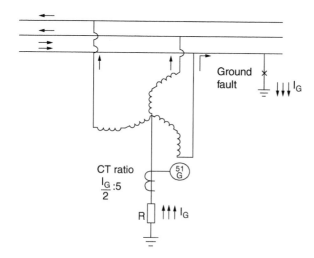

Figure 5.12 IEEE-142 (Figure 1-14(2) Low Resistance Grounding System Through ZIGZAG Grounding Transformer with Neutral Sensing Current Relay for Marine Ungrounded System).

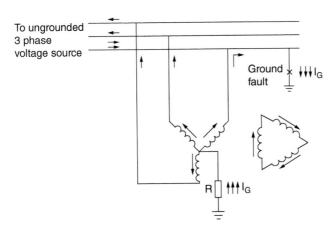

Figure 5.13 IEEE-142 (Figure 1-15(A) Wye-Delta Grounding Transformer Showing Current Flow For Marine Ungrounded System).

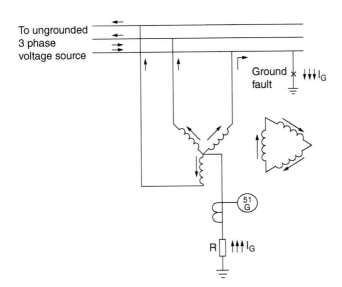

Figure 5.14 IEEE-142 (Figure 1-15(B) Low Resistance Grounding Of System Through Wye-Delta Grounding Transformer With Neutral Sensing Current Relay For Marine Ungrounded System).

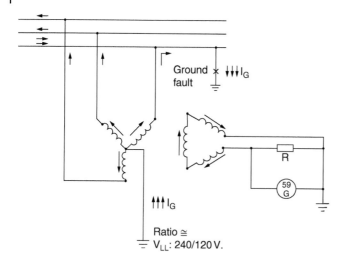

Figure 5.15 Typical Shipboard Ground Detection System- IEEE-142 (Figue 1-16) High Resistance Grounding Through A WYE-Broken Delta Grounding Transformer With Neutral Sensing Voltage Relay.

5.17 Explanations of Ground Detection System (IEEE-142 Base Figure Numbers in this Text)

A wye–delta connected three-phase transformer or transformer bank can also be utilized for system grounding. As in the case of the zigzag transformer, it can be used for effective grounding or to accomplish resistance-type grounding of an existing ungrounded system.

The delta connection must be closed to provide a path for the zero-sequence current, and the delta voltage rating is selected for any standard value. A resistor inserted between the primary neutral and ground, as shown, provides a means for limiting ground-fault current to a level satisfying the criteria for resistance-grounded systems. For this arrangement, the voltage rating of the wye winding need not be greater than the normal line-to-neutral system voltage. A neutral sensing current relay, 51G, is shown for detection of a single line-to-ground fault.

For high-resistance grounding it is sometimes more practical or economical, as illustrated in Figure 5.15, to apply the limiting resistor in the secondary delta connection. For this configuration the grounding bank must consist of three single-phase transformers with the primary wye neutral connected directly to ground. The secondary delta is closed through a resistor that effectively limits the primary ground-fault current to the desired low level. For this alternative application, the voltage rating of each of the transformer windings forming the wye primary should not be less than the system line-to-line voltage.

The rating of a three-phase grounding transformer or bank, in (kVA), is equal to the rated line-to-neutral voltage (120 V-LN) in kilovolts times the rated neutral current (see *Electrical Transmission and Distribution Reference Book*). Most grounding transformers are designed to carry their rated current for a limited time only, such as 10 s or 1 min. Consequently, they are much smaller in size than an ordinary three-phase continuously rated transformer with the same rating.

5.17.1 Transformer kVA Calculation

It is generally desirable to connect a grounding transformer directly to the main bus of a power system, without intervening circuit breakers or fuses, to prevent the transformer from being inadvertently taken out of service by the operation of the intervening devices. (In this case the transformer is considered part of the bus and is protected by the relaying applied for bus protection.)

Alternatively, the grounding transformer should be served by a dedicated feeder circuit breaker, or connected between the main transformer and the main switchgear.

The resistor tables Tables 5.1 and 5.2 below are for 10 seconds duty only. For 60 seconds duty and extended duty refer to manufacture schedule.

5.17.2 Grounding at Points Other Than System Neutral

In some cases, low-voltage systems (600 V and below) are grounded at some point other than the system neutral to obtain a grounded electrical system. This is done because delta transformer connections do not provide access to the three-phase system neutral. Two systems are in general use. See Figure 5.16.

Table 5.1 High Resistance Grounding-Resistor Selection.

230 L-L and 136 L-N System

Initial Amps	resistance value–ohms (10 seconds rating)		Remarks
	Calculated	Nominal	
1	139	133	
2	69.5	69	
3	46.3	46	
4	34.8	35	
5	27.8	29	

480 L-L and 277 L-N System

Initial Amps	resistance value–ohms (10 Seconds rating)		Remarks
	Calculated	Nominal	
1	277	276	
2	139	133	
3	92	92	
4	69	69	
5	55	54	

4160 L-L and 2400 L-N System

Initial Amps	resistance value–ohms (10 seconds rating)		Remarks
	Calculated	Nominal	
1	2400		
2	1200		
3	800		
4	600		
5	480		

7200 L-L and 4160 L-N System

Initial Amps	resistance value–ohms (10 seconds rating)		Remarks
	Calculated	Nominal	
1	4160		
2			
3			
4			
5			

Table 5.2 Low Resistance Grounding-Resistor Selection Guide.

230 L-L and 136 L-N System

Initial Amps	resistance value-ohms (10 seconds rating)		Remarks
	Calculated	Nominal	
100	1.39	1.33	
200	0.695	0.69	
300	0.463	0.46	
400	0.348	0.35	
500	0.278	0.29	

480 L-L and 277 L-N System

Initial Amps	resistance value-ohms (10 seconds rating)		Remarks
	Calculated	Nominal	
100	2.77	2.76	
200	1.39	1.33	
300	0.92	0.92	
400	0.69	0.69	
500	0.55	0.54	

4160 L-L and 2400 L-N System

Initial Amps	resistance value-ohms (10 seconds rating)		Remarks
	Calculated	Nominal	
25	96	96.6	
50	48	47.5	
100	24	24	
200	12	12	
300	8	7.82	

7200 L-L and 4160 L-N System

Initial Amps	resistance value-ohms (10 seconds rating)		Remarks
	Calculated	Nominal	
25	166.4	165	
50	83.2	85	
100	41.6	41.5	
200	20.8	20.8	
300	13.87	13.8	

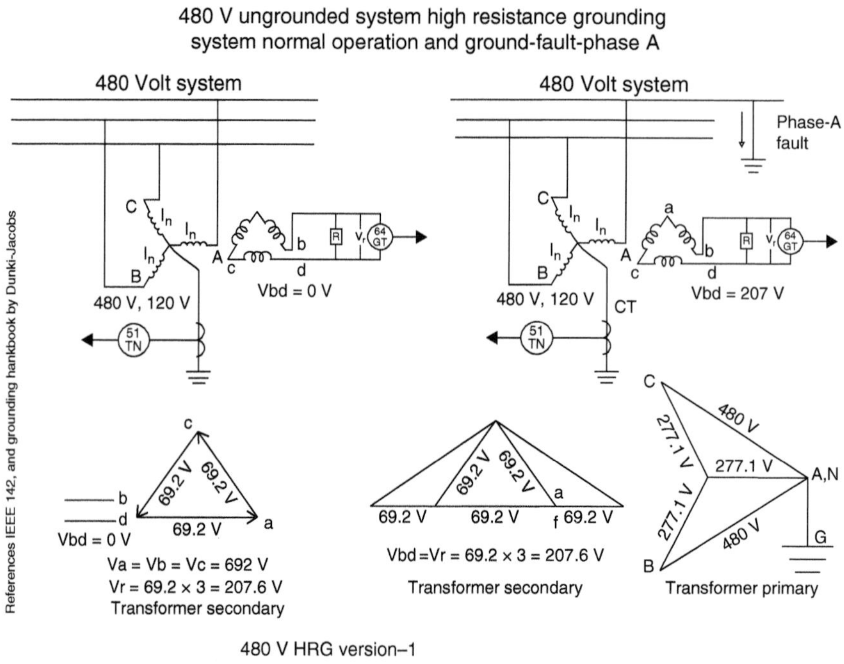

480 V ungrounded system high resistance grounding
system normal operation and ground-fault-phase A

480 V HRG version–1

Figure 5.16 480 V System With HRG Ground Detection – Normal and Faulted Condition.

5.18 The Rating And Testing Neutral Grounding Resistors- IEEE-32 Standard

Rating and Testing Neutral Grounding Resistors- IEEE-32-1972 Standards
IEEE-32 is the standard used for rating and testing neutral grounding resistors. The most important parameters to consider from the IEEE-32 are: the allowable temperature rises of the element for different "on" times; the applied potential tests; the dielectric tests, and the resistance tolerance tests that are required. Post Glover Neutral Grounding Resistors are designated and built to pass all these rigorous tests.

• Time Rating

IEEE-32 specifies standing time ratings for Neutral Grounding Resistors (NGRs) with permissible temperature rises above 30°C ambient as shown in Table 3.

Time ratings indicate the time the grounding resistor can operate under fault conditions without exceeding the temperature rises.

- 10-Second Rating

  This rating is applied on NGRs that are used with a protective relay to prevent damage to both the NGR and the protected equipment. The relay must clear the fault within 10 seconds.

- One-Minute Rating

  One NGR is often used to limit ground current on several outgoing feeders. This reduces equipment damage, limits voltage rise and improves voltage regulation. Since simultaneous grounds could occur in rapid succession on different feeders, a 10-second rating is not satisfactory. The one-minute rating is applied.

- Ten-Minute Rating

  This rating is used infrequently. Some engineers specify a 10-minute rating to provide an added margin of safety. There is, however, a corresponding increase in cost.

- Extended-Time Rating

  This is applied where a ground fault is permitted to persist for longer than 10 minutes, and where the NGR will not operate at its temperature rise for more than an average of 90 days per year.

- Steady-State Rating

  This rating applies where the NGR is expected to be operating under ground fault conditions for more than an average of 90 days per year and/ or it is desirable to keep the temperature rise below 385 °C.

  Tests

  An applied potential test (HI-POT) is required to test the insulation of the complete assembly (or sections thereof).

  For 600 volts or less, the applied potential test is equal to twice the rated voltage of the assembly (or section) plus 1,000 Volts.

  For ratings above 600 Volts, the applied potential test is equal to 2.25 times the rated voltage, plus 2,000 Volts.

  The resistance tolerance test allows plus or minus 10 percent of the rated resistance value.

IEEE-32 Table for Typical Time Rating and Permissible Temp Rise for Neutral Grounding Resistors

Item	Time Rating	Permissible Temp Rise (Above 30 °C)	Remarks
1	Short time for ten seconds	760 °C	
2	Short time for one minute	760 °C	
3	Short time for ten minutes	610 °C	
4	Continuous (steady state)	385 °C	

1) A zigzag transformer has no secondary winding and is designed to provide a low impedance path for the zero sequence current flow. During line to ground fault condition, the zero sequence current can flow into the ground at the point of fault and back through the neutral to the grounding transformer. The impedance of the zigzag transformer to balanced three-phase voltages is relatively high, therefore, when there is no fault on the system, a small magnetizing current flows in the windings.

2) A zigzag transformer with resistance ground is normally designed for short time rating 10 seconds to 60 seconds. Consequently, the grounding transformer is much smaller in size than the one with continuously rated transformer with the same rating.

6

Shipboard Power Quality and VFD Effect

Shipboard power generation and distribution system design and development must be in accordance with the applicable rules and regulations to ensure that power source characteristics are suitable for the intended ship service applications. The shipboard power systems are classified in many different categories such as type-I, type-II, emergency, essential, non-essential, vital, non-vital, sensitive, non-sensitive, etc. The commercial ship power characteristics started with MIL-STD-1399 with voltage, frequency tolerances as well as total harmonic distortion limits. The ship power system with variable frequency drive generates electric noise in many different forms deteriorating the power quality in the ship service power distribution system.

The variable frequency drives with various configuration such as six pulse drive, 12 pulse drive, 18 pulse drive, Active front end, Pulse width modulation and many others generate many different levels of harmonics. These harmonics are often much higher than the regulations allow. There are many systems available to deal with harmonic to bring down to acceptable levels. In some cases the total harmonic levels are also manipulated to a higher level giving understanding that higher level of harmonics will have no effect on the overall system.

The harmonic level may be acceptable from the VFD equipment supplier, but detrimental to the shipboard power distribution system as well as other electrical equipment in use.

The shipboard power system harmonic noise in the electrical system requires better understanding of the harmonic generation, propagation, and power system contamination issues so that necessary steps are taken to bring the power system THD level to the acceptable level.

Therefore the harmonic noise management to meet the regulatory body requirements are used such as low pass power system filter, band pass power system filter, hybrid power system filter, active power system filter and combinations thereof.

The variable frequency drive selection is very complex due to fact that shipboard islanded system-level applications are often ignored during early stages

VFD Challenges for Shipboard Electrical Power System Design, First Edition.
Mohammed M. Islam.
© 2020 by The Institute of Electrical and Electronics Engineers, Inc.
Published 2020 by John Wiley & Sons, Inc.

of design. The shipboard power distribution systems require systematic power quality calculation and system level impact analysis related verification so that system level power quality is not compromised.

There are applications which must be re-evaluated and necessary actions must be taken in view of the operational requirements as well as electrical safety of ships.

6.1 Motor Starting Current With Various Starters

The motor starting current requirement can be 100% to 1200% of the full load current, which must be taken under consideration for power system design and generator loading. See Table 6.1 for details.

The industrial distribution system is basically from the distribution transformers. The shipboard power is from dedicated generators. The utility based distribution bus impedance is lower than the shipboard islanded distribution bus with dedicated generator related impedance.

6.2 Harmonics Requirements: IEEE-519-1992 versus IEEE-519-2014

The IEEE-519-1992 has been superseded by IEEE-519-2014 edition. However, the IEEE-519-2014 is a simplified version of 1992 version. For shipboard application the IEEE-519 may not provide enough guidelines for Voltage and current harmonic calculation and mitigation.

Table 6.1 Motor Starter Inrush Current Requirements.

Motor Starter Type	Motor Starting Current (% Of Full Load Current)	Remarks
Across the line starter	600–800%	Significant starting voltage drop issue on the main bus, 8 times
Autotransformer starter	400–500%	Moderate starting voltage drop issue on the main bus, 5 times
Wye/delta starter	200–275%	Minimum Starting voltage drop issue on the main bus, 3 times
Solid state soft starter	200%	Minimum starting voltage drop issue on the main bus, 2 times
VFD/ASD	100%	1) No voltage drop issue on the main bus. 2) Most reliable application for shipboard islanded system

The IEEE-519-2014 version also increased the voltage harmonics from 5% to 8% at the shipboard voltage level.

American Bureau Shipping Guideline Notes on "Control of Harmonics in Electrical Power System" Publication _150 EI Harmonics provides excellent tutorial for shipboard and offshore platform power generation and distribution related voltage and current harmonics calculations in different configurations of harmonic generating loads. The document also outlines details of the harmonic mitigating means to comply with the harmonic limitation requirements such as the application of reactors, passive filters, wideband filters, active filters and hybrid filters at six pulse, 12 pulse, 18 pulse configuration.

The VFD or ASD shipboard application is very complex design and development process. The harmonic generation as well as mitigation must be phased in at the preliminary design phase by the subject matter experts having jurisdiction in this field. The shipbuilder, the designer, VFD supplier and harmonic mitigating experts must work together to establish proper harmonic management system. See Table 6.2.

The following VFD/ASD applications are analyzed for design considerations:

1) 6 Pulse drive
2) 6 pulse drive with reactors
3) 6 pulse drive with passive filters

Table 6.2 Typical Voltage Harmonic Orders for VFD.

Drives	Voltage Harmonic Order								Up to 49th	Remarks
	5th	7th	11th	13th	17th	19th	23rd	25th		
6-Pulse	0.175	0.11	0.045	0.029	0.015	0.009	0.008	0.007	≥27%	Manage significant harmonics 5th and 7th
12-Pulse	0.026	0.016	0.073	0.057	0.002	0.001	0.02	0.016	≥11%	Manage significant harmonics 11th and 13th
18-Pulse	0.021	0.011	0.007	0.005	0.045	0.039	0.005	0.003	≥6.6%	Manage 17th and 19th
24-Pulse									<5%	
AFE-PWM	0.037	0.005	0.001	0.019	0.022	0.015	0.004	0.0035	<5%	1. Generates EMI & RFI. 2. Must consider harmonic calculation up to 99th

4) 6 pulse drive with reactors and passive filters
5) 12 Pulse drive
6) 12 pulse drive with reactors
7) 12 pulse drive with passive filters
8) 12 pulse drive with reactors and passive filters
9) 18 Pulse drive with reactors and passive filters
10) Active front end (AFE) PWM drive

6.3   Solid State Devices Carrier Frequency

Points to be checked are as follows:

a) SCR (Silicon controlled rectifier) – What is the rise time of the variable speed drive's output IGBTs? 2 kHz to 8 kHz can operate in the 250 to 500 Hz range, (4 to 8 times fundamentals).
b) BJTs (Bipolar Junction Transistor) can operate in the 1 kHz to 2 kHZ range, (16 to 32 times fundamentals).
c) Insulated gate bipolar transistors (IGBT) for the inverter section. IGBTs can turn on and off at a much higher frequency 2 to 20 kHz. (30 to 160 times fundamentals).
d) Typical AFE (Active front end with IGBT) switching frequency 3600 Hz (60 times fundamentals).
e) IGCT: The IGCT is optimized for low conduction losses. Its typical turn-on/off switching frequency is in the range of 500 hertz. See Table 6.3.

PWM Drive: The motor electrical frequency is modulated at a much higher rate using pulse width modulation techniques at the range of 2 kHz to 30 kHz, which is called PWM carrier frequency. The carrier frequency at that range (2 kHz to 3 kHz) will have very high dv/dt producing electro-magnetic interference (EMI) or RFI Radio Frequency interference.

These drives generate harmonics creating many noise issues in the power system. Some of the noise implications are very complex. Therefore, it is very

Table 6.3 Solid State Device Sample Carrier Frequency.

Solid state device type	Carrier frequency	Explanation
SCR	250 Hz to 500 Hz	4 to 8 time the fundamentals
BJT	I kHz to 2 kHz	8 to 16 times the fundamentals
IGBT	2 kHz to 20 kHz	16 to 160 times the fundamentals
PWM	2 kHz to 30 kHz	16 to 240 times the fundamentals

important to understand various noise issues, such as origination, noise effects, and then manage noise within reason level to be in compliance with the requirements. The VFD-related noise can generate heat directly degrading equipment and premature failures. The shipboard electrical equipment usually design for twenty to thirty years life depending on the application. However, recently electrical equipment failure trend has changed, leading to premature failure. The shipboard electrical systems are categorized as critical and noncritical. Most of the propulsion related equipment are in critical category. There are requirements to provide redundant system to be in compliance with the requirements.

Many propulsion auxiliaries are also being provided with VFD. The shipboard power distribution is mostly radial bus with some built-in isolation capability. The isolation is to ensure power segregation in case of major failures in one section of the switchboard. Therefore, the VFD-related noise issues propagates entire power system as there is no requirement nor any provision for system level isolation. However, the requirements are to maintain the THD level at the distribution switchboard.

Electrical propulsion is the most acceptable solution comparing with other traditional propulsions. The electrical propulsion also is high power requirement due to propulsion power requirements which can be from 15 megawatts to 40 megawatts. The USCG Healy Ice breaker propulsion power is approximately 20 megawatts. Due to the propulsion power generation requirement, the power generation is 4160 V to 15 kV. There are other smaller propulsion requirements such as thrusters. There can as many as six thrusters as a requirement by applications.

To accommodate the VFDs around 1-3 MW, 690 VAC bus has been introduced. Therefore, all VFD loads and auxiliary load distribution is with step down transformers.

The harmonic calculation requirements are to establish the point of common coupling (PCC). For shipboard applications, the PCCs can the propulsion switchboard bus and the power transformer secondary. The total harmonic distortion (THD) requirement is not to exceed 5%. The THD can be over 5% if the equipment at that system selected to withstand the available harmonic. But in no case can the total harmonic be over 8%.

The following is shipboard VFD application in categories in application to comply with the THD limit:

1) 18 Pulse drive with phase shift transformers
2) 12 pulse drive with phase shift transformers and filters
3) 6 Pulse drive with filters
4) Active front end drive
5) PWM drive.

The 18 pulse drive is robust, expensive and space demanding due to three transformer requirements. However, 18 pulse drive usually meets the harmonic requirements of maximum THD 5%.

The 6 Pulse drive is the least expensive with minimum space requirements. However, the 6 Pulse drive system may have high THD level. Therefore 6 Pulse drive are used with filters to minimize the THD level to meet requirements. There are different types of filters available such as passive filter and active filter. The 6 Pulse drive with active filter and reactor has proven to be the best design comparing with the size, cost and reliability.

6.4   MIL-STD-1399 THD Requirements

For shipboard power system characteristics of nominal voltage 440V and less see table 1.1. For shipboard power system characteristics of nominal voltage 5kV to 15 kV see Table 6.4.

6.5   IEEE-519 THD Requirements
(1992 and 2014 Version)

For shipboard power generation and distribution related harmonic management both voltage and current harmonic limits must be established. At various power distribution levels both voltage and current harmonic level must be established and then must be accumulated for system level consideration. This can only be done when proper PCC is determined during the system design and development. The harmonic levels should be calculated at every PCC and then accumulative effect be established to establish system level distortion as required by Mil Std - 1399. If the harmonic levels are calculated to be higher than the requirement outlined in a dedicated system, the distribution at the coupling point must utilize proper equipment to withstand that particular harmonic level.

At the ship service user level such as 450 V, 120 V system, the user contributes to the current and voltage harmonics, measurement must be taken, and appropriate measure must be taken to limit propagation of harmonics to the system and to maintain harmonics within limits. In general, each ship must be provided with calculated and measured harmonic level at the time of delivery of the ship.

At the sub-system level, it is not permissible to control or alter the system impedance characteristics to reduce voltage distortion and should not add passive equipment that affects the impedance characteristic in a way such that voltage distortions are increased. This adjustment is allowed by the authority having jurisdiction for this action.

In addition to the MIL-STD-1399 THD requirements, IEEE-519 is the recommended practice for all commercial and industrial applications. The IEEE-519-1992 provides detailed understanding of harmonics with examples

Table 6.4 MIL-STD-1399-680 Power System Characteristics.

Table 7.5 MIL-STD-1399-680: Table-II Electrical Power System Characteristics at the Interface (High Voltage For Shipboard Application) (Partial Extract)

Characteristics	5 kV Class	8.7 kV Class	15 kV Class
1) Nominal frequency	60 Hz	60 Hz	60 Hz
2) Frequency tolerances	±3%	±3%	±3%
3) Frequency modulation	½%	½%	½%
4) Frequency transient	±4%	±4%	±4%
5) The worst-case frequency excursion from nominal frequency resulting from item 2, item 3, and item 4 combined, except under emergency conditions.	±5.5%	±5.5%	±5.5%
6) Recovery time from Items 4 or 5	2 SEC	2 SEC	2 SEC
7) Nominal user voltage	4.16 kV rms	6.6 kV rms	11 kV rms, 13.8 kV rms
8) Line voltage unbalance	3%	3%	3%
9a) Average of the three line-to-line voltages	±5%	±5%	±5%
9b) Any one line-to-line voltage, including item 8 and 9a	±7%	±7%	±7%
10) Voltage modulation	2%	2%	2%
11) The maximum departure voltage resulting from item 8, 9a, 9b and 10 combined, except under transient or emergency conditions.	±6%	±6%	±6%
12) Voltage transient tolerance	±16%	±16%	±16%
13) Worst case voltage excursion from nominal user voltage resulting from items item 8, 9a, 9b, 10 and 12 combined, except under transient or emergency conditions.	±20%	±20%	±20%
14) Recovery time from items 12 or 13	2 sec	2 sec	2 sec
15) Voltage spike	60 kV peak	75 kV peak	90 kV peak
16) Maximum total harmonic distortion	5%	5%	5%
17) Maximum single harmonic	3%	3%	3%
18) Maximum deviation factor	5%	5%	5%

for the calculation of the harmonic level. However, the IEEE-519-1992 has been superseded by IEEE-519-2014, which provides some additional clarification, but there is no clarification of the requirements and no examples to follow. Therefore, the user of the IEEE-519 for shipboard power system design and development must understand the differences between the two versions of IEEE-519. Some of the differences are given in tables 6.5 and 6.6.

6.5.1 Total Harmonic Distortion (THD)

Table 6.5 THD Definition Comparison (IEEE-519-1992 AND 2014).

Total Harmonic Distortion (THD)		
THD-IEEE-519-2014	**TDD-IEEE-519-1992**	**Remarks**
TOTAL HARMONIC DISTORTION (THD): The ratio of the root mean square of the harmonic content, considering harmonic components up to the 50th order and specifically excluding inter-harmonics, expressed as a percent of the fundamental. Harmonic components of order greater than 50 may be included when necessary. (IEEE-519-2014)	TOTAL HARMONIC DISTORTION (THD): This term has come into common usage to define either voltage or current "distortion factor". Where DISTORTION FACTOR is the ratio of the root mean square of the harmonic contents to the root mean square value of the fundamental quantity, expressed in percentage of fundamentals.	IEEE-519-1992 requirement was voltage or current. IEEE-519-2014 does not specifically states as voltage or current, instead calls for harmonic contents

6.5.2 Total Demand Distortion (TDD)-Current Harmonics

Table 6.6 Total Harmonic Distortion (THD) Voltage Limit Comparison IEEE-519-1992 Version and IEEE-519-2014 Version).

Bus Voltage @ PCC	Individual Voltage Harmonic(%) PER 519-2014	Individual Voltage Harmonic(%) PER 519-1992	Total Voltage Harmonic Distortion -THD(%) PER 519-2014	Total Voltage Harmonic Distortion = THD(%) PER 519-1992	Remarks
V ≤ 1.0 kV	5	none	8	none	IEEE-519-2014 THD limit of 8% does not agree with established standards for shipboard power system.
1 kV < V ≤ 69 kV	3	3	5	5	Does agree with established standards for shipboard For 519-1992 the voltage range was 69 kV and below power system.

6.6 Current Harmonic

Table 6.7 Total Demand Distortion (TDD) Current Limit (Extract From IEEE-519-2014 Table-2).

I_{SC}/I_L	$3 \leq h < 11$	$11 \leq h < 17$	$17 \leq h < 23$	$23 \leq h < 35$	$35 \leq h < 50$	TDD	Remarks
<20	4.0	2.0	1.5	0.6	0.3	5.0	See note-a
20 < 50	7.0	3.5	2.5	1.0	1.5	8.0	
50 < 100	10.0	4.5	4.0	1.5	0.7	12.0	
100 < 1000	12.0	5.5	5.0	2.0	1.0	15.0	

Table 6.8 Total Harmonic Distortion (THD) Voltage Limit Comparison Between Different Drives.

Attributes	6-Pulse	6-Pulse with Filter	12 Pulse	18 Pulse	Active Front End
Typical THD_V at the drive input terminal	@30–40%	@5–8%	5–12%	5–8%	3–4%
Weight	100%	150%		240%	200%
Dimension	100%	140%		160%	170%

These are approximate values for information only. These values may vary widely from application to application.
Active front end unit is the most expensive.
Drive transformers for 12 pulse and 18 pulse drive will have major weight and volume impact due to transformers.

6.7 Harmonic Numbering

Use this equation: $h = kq \pm 1$. See Table 6.9 below.

6.8 DNV Regulation - Harmonic Distortion

Guidelines are as below:

a) Equipment producing transient voltage, frequency and current variations shall not cause malfunction of other equipment on board, neither by conduction, induction, or radiation.

Table 6.9 Harmonic Contents for Various VFDs.

Harmonic numbering	6 pulse drive	12 pulse drive	18 pulse drive	Remarks
3				
5	5			
7	7			
9				
11	11	11		
13	13	13		
15				
17	17		17	
19	19		19	
21				
23		23		
25		25		
27				
29				
31				
33				
35		35	35	
37		37	37	
39				
41				
43				
45				
47				
49			53	
51			55	

b) In distribution systems the acceptance limits for voltage harmonic distortion shall correspond to IEC 61000-2-4 Class 2. (IEC 61000-2-4 Class 2 implies that the total voltage harmonic distortion shall not exceed 8%.). In addition, no single order harmonic shall exceed 5%.

c) The total harmonic distortion may exceed the values given in (b) under the condition that all consumers and distribution equipment subjected to the increased distortion level have been designed to withstand the actual levels. The system and components ability to withstand the actual levels shall be documented.

d) When filters are used for limitation of harmonic distortion, special precautions shall be taken so that load shedding or tripping of consumers, or phase back of converters, do not cause transient voltages in the system in excess of the requirements. The generators shall operate within their design limits also with capacitive loading. The distribution system shall operate within its design limits, also when parts of the filters are tripped, or when the configuration of the system changes.

Guidance note: The following effects should be considered when designing for higher harmonic distortion in (c):

- additional heat losses in machines, transformers, coils of switchgear and control gear
- additional heat losses in capacitors for example in compensated fluorescent lighting
- resonance effects in the network
- functioning of instruments and control systems subjected to the distortion
- distortion of the accuracy of measuring instruments and protective gear (relays)

B) Remarks for DNV Harmonic requirements: Failure mode and effect analysis shall be performed at the shipboard system level for any filter application to ensure the cause and effect to the system as to the filter malfunction. If necessary, shipboard operational adjustment must be made to protect equipment and system.

6.9   Choice Of 18 Pulse Drive Versus 6 Pulse Drive With Active Harmonic Filter

The 18 pulse drive system is a better choice for harmonic management. However, th pros and cons of using 18 pulse drive versus six pulse drive with filters, refer to the Table 6.10.

6.10   Typical Calculation of Total Harmonic Distortion and Filter Applications

Typical shipboard power system with Variable frequency drive (VFD) or Adjustable speed drive (ASD) is shown in Figure 4.1 consisting of 6600 V distribution bus, 690 V distribution bus and 480 V distribution bus. This example is shown is to calculate the harmonic distortion level and then apply various harmonic mitigating features to reduce the harmonic to IEEE-519 acceptable level.

Table 6.10 Filter Comparison Table.

18 Pulse drive	Active Harmonic Filter (AHF)	Remarks (Use of 6-pulse drive with AHF)
1. Will meet IEEE-519 Harmonic requirements. Drive consists of three-phase shift transformers, transformer protection, and insulation heat management. The transformer may require pre-magnetization depending on the size. Transformer efficiency is low. For a large propulsion drive transformer heat dissipation in the transformer room is very high. Transformer power cable primary and secondary adds tremendous amount of weight. Shipboard transformer cable routing is a big challenge due to so many cables involved. Shipboard cable termination, particularly for VFD application is also complex.	Will meet IEEE-519 Harmonic requirements. The AHF can be integrated reactors to keep the size of the AHF smaller. The reactors are robust and long lasting without any maintenance	Six pulse drive is such simpler than 18 Pulse drive. 6 Pulse drive, part replacement, repair, maintenance is less complicated than the 18 pulse drive
2 Will not tolerate system voltage imbalance	Will tolerate system voltage imbalance	
3 Transformers are very heavy and space demanding	AHF is not heavy, nor space demanding comparing with transformers.	
4 Transformer cooling is challenging leading to more cooling equipment demanding additional space and weight	Do not know the cooling issues for AHF	
	The AHF is modularized design capable of multiple parallel connections. This contributes to easier size selection for ships due to head room limitations.	
	Due to application of intelligent electronics and IGBTs the system can deliver corrective amount of current and manage harmonic at the acceptable level.	
The transformer cost will increase due to the increase in cost of copper.	The AHF cost will come down as the power electronic use increases and the rating increases.	

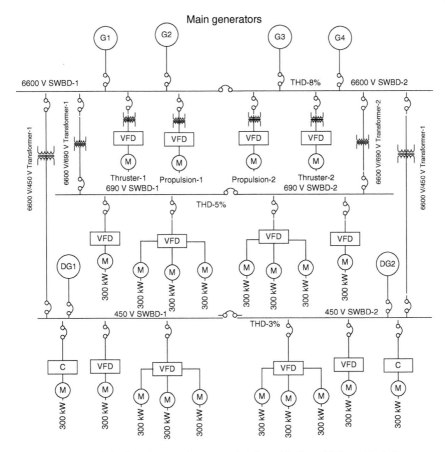

Figure 6.1 Typical Shipboard Power System With Adjustable Speed Drives –Typical.

To select the correct harmonic filters for the system shown in Figure 4.1, the two 690 V busses are analyzed separately. Additionally, since they are both main-tie-main systems, the system is modeled as two individual circuits; one on each side of the tie. The 690 V bus SWBD-1 is supplied by a 2,500 kVA, 6.6 kV to 690 V transformer with 5.75% impedance and feeds (1) 300 kW 6-pulse VFD and (1) 900 kW active front end drive connected to 3 ea. 300 kW motors. The 6-pulse drive is assumed to have 4% internal DC impedance, typical for a variable frequency drive. See Figures 6.1 to 6.4.

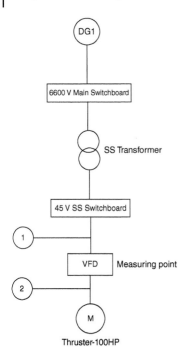

Figure 6.2 Example Of VFD Current Calculation.

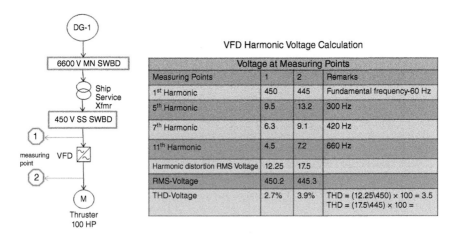

VFD Harmonic Voltage Calculation

Voltage at Measuring Points			
Measuring Points	1	2	Remarks
1st Harmonic	450	445	Fundamental frequency-60 Hz
5th Harmonic	9.5	13.2	300 Hz
7th Harmonic	6.3	9.1	420 Hz
11th Harmonic	4.5	7.2	660 Hz
Harmonic distortion RMS Voltage	12.25	17.5	
RMS-Voltage	450.2	445.3	
THD-Voltage	2.7%	3.9%	THD = (12.25\450) × 100 = 3.5 THD = (17.5\445) × 100 =

Figure 6.3 6600 V Generation and 480 V Distribution Voltage Harmonic Calculation.

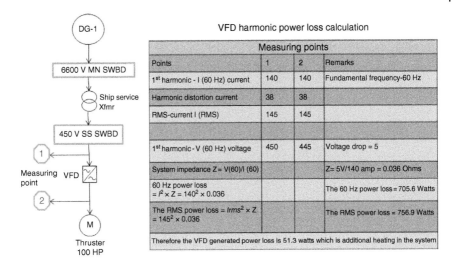

Figure 6.4 VFD Harmonic Power Loss Calculation.

7

Shipboard Power System FMEA for VFD Motor Controller

Shipboard power system design and development includes identification of vital equipment and/or system, to establish redundancy features as well as vital equipment, and to perform failure mode and effect analysis (FMEA) to prevent failure situation and then initiate recovery as regulatory body requirement under the context of safety and security of the ship and shipboard people.

The shipboard electrical system with VFD supported vital equipment falls in this category and only some critical equipment with VFD motor controller is presented, though the system may not be identified by regulations.

This book addresses both traditional design and development issues namely:

a) system design process
b) system developmental process
c) verification of the design and development for operational requirements
d) verification of the design and development for regulatory requirements
e) verification of failure mode and effects of the design
f) verification of system behavior and maintenance requirements for the operators
g) verification of training requirements for the design
h) verification of operators readiness

7.1 FMEA Analysis-Adaptation

The findings of the FMEA analysis can be validated by correct or complete design information. The findings of the FMEA analysis can be invalidated by incorrect or incomplete design information or any modification.

The operators and engineers must be aware of the consequences of the failure of any single item of equipment that they have chosen as a vital equipment

VFD Challenges for Shipboard Electrical Power System Design, First Edition.
Mohammed M. Islam.
© 2020 by The Institute of Electrical and Electronics Engineers, Inc.
Published 2020 by John Wiley & Sons, Inc.

and to operate the vessel within the worst case failure environmental limits. The vessel should be operated at all times bearing in mind the selected equipment capacity after worst-case failure.

7.2 Typical Design and Development of Power Generation to Serve Typical Ship Service Load

The Figure 7.1 is a typical shipboard power generation and distribution system delivering power to vital loads 1, 2, and 3. The main generator is delivering 6600 V power to a 6600 V switchboard. The 6600 V switchboard is delivering power to the 480 V ship service switchboard through a step down transformer. The 480 V switchboard is supplying power to a 450 V VFD motor controllers for the vital loads.

a) Figure 7.2 is a typical shipboard power generation and distribution system delivering power VFD/ASD motor controllers in the 6600 V switchboard, 690 V switchboard, and 480 V ship service switchboard. The FMEA analysis should include the vital equipment in each voltage class 6600 V, 690 V class and 480 V class VFD/ASD controllers, including the VFD generated electric noise contribution to each system and then to the overall system. Each voltage

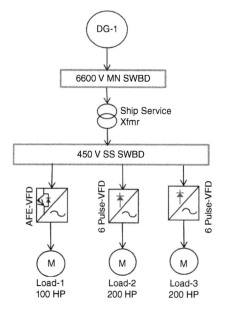

Figure 7.1 Typical Sample Electrical Plant With FMEA.

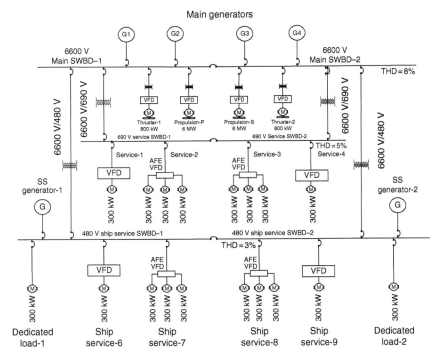

Figure 7.2 Typical Electrical Plant With Many VFD Applications With FMEA Requirements.

class must consider the harmonic limits independently and then as a complete system due to the fact that the harmonic distortion affects adversely the entire power system. Analysis must be done to ensure system level integrity as well as operational capability of the entire system.

b) For dynamic positioning (DP) class vessel, the requirements are very strict due to the fact that the thrusters, propulsion, and supporting vital auxiliaries must operate as designed with multiple redundant systems and equipment, The FMEA must address each and every situation due to DP operational criticality.

c) Figure 7.3 is a typical shipboard power generation and distribution system delivering power to a lube oil pump. Main generator is delivering 6600 V power to a 6600 V switchboard. The 6600 V switchboard is delivering power to the 480 V ship service switchboard through a 5 MVA step down transformer. The 480 V switchboard is supplying power to a 450 V vital power panel. The vital power panel is supplying power to the lube oil pump motor controller and motor.

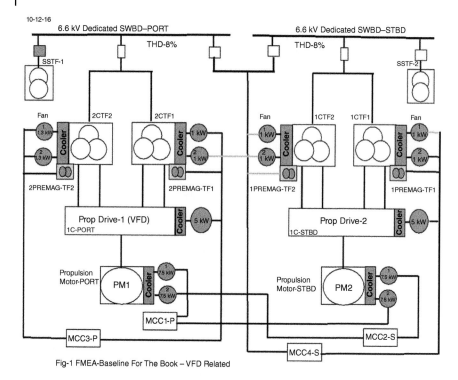

Fig-1 FMEA-Baseline For The Book – VFD Related

Figure 7.3 Typical ASD Electrical Propulsion and FMEA Requirements.

7.3 Design Verification – General

A design is to meets the fundamental requirements by contract specification, which is presentation in electrical one line diagram. The basic verification is outlined in the USCG regulations called failure mode and effect analysis (FMEA) supported by qualitative failure analysis (QFA) and design verification test procedure (DVTP).

7.4 Qualitative Failure Analysis (QFA)

The QFA must be prepared assuming the vessel is in its normal condition of operation and reflect the level of automation and manning level of the machinery plant, e.g. vessel underway in pilothouse control, all main engines in remote automatic operation, machinery space manned or unattended (depending on the vessel's manning level), and automatic power management system, if provided, is active. Checking the QFA's Failure Effects: a. Propulsion Control Systems. 46 CFR 62.35-5(e)(3).

7.4.1   Qualitative Failure Analysis (QFA) Basics

The following procedure should be performed:

a) Perform this maintenance procedure after reviewing the results from the most recent infrared thermographic inspection of the VFD controller converter cubicles
b) Secure and tag out the VFD motor controller:
    i) Verify that the VFD converter is in DRIVE SECURED mode.
    ii) Set VFD transformer feeder breaker local/remote control switch to LOCAL position at the Switchboard.
    iii) Verify that converter Transformer feeder breaker is in OPEN position, and tag out breaker.

7.4.2   Process Failure Mode and Effect Analysis –FMEA –General

The QFA must be prepared assuming the vessel is in its normal condition of operation and reflect the level of automation and manning level of the machinery plant, e.g. vessel underway in pilothouse control, all main engines in remote automatic operation, machinery space manned or unattended (depending on the vessel's manning level), and automatic power management system, if provided, is active.

Checking the QFA's Failure Effects: a. Propulsion Control Systems. 46 CFR 62.35-5(e)(3).

1) Failures of the remote propulsion control system should be failsafe, such that the preset speed and direction (as-is) of thrust is maintained, until local manual or alternate manual control is in operation, or the manual safety trip (shutdown) is activated. This is required specifically for vessels with a single propulsion plant or single propeller.
2) For a vessel with multiple and independently controlled propellers, a failure of one propulsion control system need not follow the failsafe requirements above. The failsafe options available in this case are:
    i) Force both control systems to fail "as-is." Systems respond like a vessel with a single propulsion plant.
    ii) Fail "as-is" of just the affected control system, while maintaining full control of the unaffected propulsion system.
    iii) Fail to "zero" thrust or trip of the affected propulsion system, providing partial reduction of normal propulsion capability as a result of malfunction or failure. Reduced capability should not be below that necessary for the vessel to run ahead at 7 knots or half-speed of the vessel, whichever is less, and is adequate to maintain control of the ship. This adopts the intent of the "Note" in 46 CFR 58.01- 35.

7.4.3   Qualitative Failure Analysis (QFA)-1

See Table 7.1.

7.4.4   Explanation of the Detail Design Using QFA

See Table 7.2.

7.5   Design Verification Test Procedure General – DVTP

The Design Verification Test Procedure (DVTP) document is required to be "Approved" and retained aboard the vessel. Using the DVTP document, design verification testing is required to be performed immediately after the installation of the automated equipment or before the issuance of the initial Certificate of Inspection. Final approval of the DVTP document is contingent upon satisfactory completion of onboard design verification tests in the presence of the Coast Guard. See 46 CFR 61.40-1(c), and 62.30-10(a).

Applicable to self-propelled vessels of 500 gross tons and over that are certificated under subchapters D, I, and U, and to self-propelled vessels of 100 gross tons and over that are certificated under subchapter H.

The Design Verification Test Procedure (DVTP) document is required to be "Approved" and retained aboard the vessel. Using the DVTP document, design verification testing is required to be performed immediately after the installation of the automated equipment or before the issuance of the initial Certificate of Inspection. Final approval of the DVTP document is contingent upon satisfactory completion of onboard design verification tests in the presence of the Coast Guard. See 46 CFR 61.40-1(c), and 62.30-10(a). Design verification testing is used to verify the automated vital system installations are designed, constructed and operate in accordance with the applicable requirements in 46 CFR Part 62. See 46 CFR 61.40-3.

The design verification test procedures may be incorporated with the qualitative failure analysis (QFA). See E2-18 Work Instruction. The DVTP document is a separate document from the Periodic Safety Test Procedure (PSTP) document. Both documents are required to be approved and retained aboard the vessel. See 46 CFR 61.40-1(c).

- Title 46 CFR Parts 58, 61 and 62
- Title 46 CFR Parts 111 and 112
- Navigation and Inspection Circular (NVIC) 2-89, "Guide for Electrical Installations on Merchant Vessels and Mobile Offshore Drilling Units"
- American Bureau of Shipping (ABS), "Rules for Building and Classing Vessels under 90 Meters in Length", 1996

Table 7.1 Design Verification Matrix – QFA.

Item	System	Sub-system	Failure point	Unit	Local diagnostic	Alarm point	Local/remote	Effects	Remarks
1	Main supply switchboard	Voltage	Transient	Voltage	Meter	Hi-low	Local/remote	Degrade over all operation	System level
1a		Harmonics							
1b		Frequency		Hz		Fluctuating	Local/remote	Degrade over all operation	System level
2	Prop transformer	Winding temp		C - degree	Meter	Hi point	Local/remote	Degrade over all operation	

Table 7.2 Typical (QFA) Qualitative Failure Analysis – Generation.

System: Propulsion Plant			Main Propulsion Generator			
1	2	3	4	5	6	8
No.	Function	Failure mode	Failure cause	Failure detection	Corrective action	Failure evaluation
1	Generation	Short circuit		Alarm	No	Max. 80% total propulsion
2	Generation	Overcurrent		Alarm	No	Max. 80% total propulsion
3	Generation	Reverse power		Alarm	No	Max. 80% total propulsion
4	Generation	Under-voltage	actual value failure	Alarm	No	Max. 80% total propulsion
5	Generation	Over-voltage	actual value failure	Alarm	No	Max. 80% total propulsion

- Safety Of Life at Sea (SOLAS), Consolidated Editions, 1997, Chapter II-1, Part D
- MSC Procedure E2-1, Vital System Automation Work Instruction

These guidelines were developed by the Marine Safety Center staff as an aid in the preparation and review of vessel plans and submissions. They were developed to supplement existing guidance. They are not intended to substitute or replace laws, regulations, or other official Coast Guard policy documents. The responsibility to demonstrate compliance with all applicable laws and regulations still rests with the plan submitter. The Coast Guard and the U.S. Department of Transportation expressly disclaim liability resulting from the use of this document.

If you have any questions or comments concerning this document, please contact the Marine Safety Center by e-mail or phone. Please refer to the Procedure Number: E2-18:

- QFA General Acceptance Criteria:
  Fail safe state must be evaluated for each subsystem,/system or vessel to determine the least critical consequence. Lowest level of system component failure to be considered is: "easily replaceable component." 46 CFR 62.30-1(a).

  The DVTP document, if submitted separately with the OFA document, must include the following QFA document information: – Component Failure Considered are "Failure Effects" and "Failure Detection".

Examine the test instructions to insure that they closely or realistically simulate the failure of only the failed component of each of the failures considered in the failure analysis. For example: A PLC power supply module failure may be tested by removing the fuse to the power supply module, but a CPU failure (served by the same power supply module), should not be tested using the same power supply fuse, as it is desired that the power supply remain in operation, with just the CPU failing.

Test instructions should be prepared as if the vessel is underway, in pilothouse automatic pilothouse control, various machinery automation in normal underway mode of operation, and the engine room manned to the manning level design of the machinery plant.

Design verification testing using the failures considered in the QFA, the vital system automation installation, although supplied by various manufacturers, should function as an integrated system, i.e., various automated systems, although supplied by separate manufacturers, may be used to monitor the operational integrity of other systems and provide failure alarms.

7.5.1 Example of Propulsion Plant-(DVTP) Design Verification Test Procedure

See Table 7.3.

Table 7.3 Typical DVTP for Propulsion Plant Power Generation.

Propulsion Plant-(DVTP)			Main Propulsion Generator-Power System			
1	2	3	4	5	6	8
No.	Failure description	System effect due to failure	Failure detection/ indication	Effect on overall ship performance due to failure	Corrective action	Failure evaluation (Remaining propulsion)
1	Generation	Short circuit		Alarm	No	Max. 80% propulsion
2	Generation	Overcurrent		Alarm	No	Max. 80% propulsion
3	Generation	Reverse power		Alarm	No	Max. 80% propulsion
4	Generation	Under voltage	Actual value deviation	Alarm	No	Max. 80% propulsion
5	Generation	Overvoltage	Actual value deviation	Alarm	No	Max. 80% propulsion

7.6 Automation – FMEA Matters

Programmable control or alarm system logic must not be altered after satisfactory completion of Design Verification Tests without the approval of the cognizant Officer in Charge, Marine Inspection. This comment is usually included in the approval letter of the DVTP document to ensure the cognizant OCMI and the ship's owner are aware of the requirements. See 46 CFR 62.25-25(a). This means that the DVTP document is only used during the initial issuance of the vessel's certificate of inspection or immediately after the installation of the automated equipment, and when the installed automated equipment is upgraded or altered. For the periodic safety test procedure (PSTP) document, periodic safety testing is conducted at periodic intervals specified by the regulation.

7.7 VFD/ASD Motor Controller FMEA Initiative

7.7.1 6600 V Supply Transformer

The propulsion drive VFD power is supplied from dedicated 6.6 kV/7200 V transformer. The transformer has an auxiliary service that is supplied from the 480 V switchboard. The VFD/ASD monitors the transformer for faults. It starts and stops the transformer auxiliary services. There are transformers with phase-shift winding, which provides a manageable degree of harmonic distortion cancellation depending on loads. Other transformers also provide some galvanic isolation. However, the transformer is a vital piece of equipment for the propulsion system. Therefore, FMEA should be established for the propulsion transformers.

7.7.2 690 V Supply Transformer

There are VFDs power is supplied from, 6.6 kV/690 V transformer. The transformer has an auxiliary service that is supplied from the 480 V ship service switchboard. These transformers provide galvanic isolation and the distribution is called dedicated bus. The dedicated bus total harmonic distortion is different from other distribution systems. The FMEA should be established for those transformers.

7.7.3 480 V Supply Transformer

There are VFDs where power is supplied from a 6.6 kV/480 V transformer. These transformers provide galvanic isolation.

However, there are VFDs where power is supplied from 480 V ship service distribution system. Usually, there is no galvanic isolation for this type of system. All ship service loads and low voltage loads are also from the same distribution system. There is no prohibition of using VFD motor controllers from this type of distribution. However, the harmonic distortion can easily contaminate the entire distribution system. Therefore, it is strongly recommended that FMEA be performed, in view of harmonic distortion limits, as well as other precautionary measures to establish that the power system within the boundary of all required operating characteristics such as harmonics, transients, etc.

7.7.4 VSD/ASD Motor Controller FMEA

The propulsion system ASD controls the propeller speed by varying the motor speed. It changes the rotation of the motor to change the direction of propulsion thrust. The ASD may be equipped with separately mounted air cooled breaking resistors for faster propeller reversal. The propulsion ASD related FMEA is a requirement of the regulatory body to ensure proper operation as well as redundancy requirements.

7.8 Ship Service Generator

Ship service generator control block diagram is shown in Figure 7.4 is to show different components of the ship service generator, prime mover, control, and protection. All of these are mandatory for the development of FMEA.

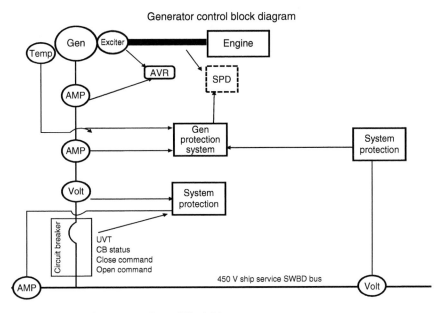

Figure 7.4 Typical Generator Control Block Diagram.

8

Shipboard VFD Cable Selection, Installation, and Termination

IEEE recommendations for shipboard electrical cable construction, application and installation are in IEEE-45, IEEE-1580, and IEEE-45.8.

US Navy cable construction and performance specifications are in MIL Specifications, such as MIL-C-24643, MIL-C-24640, MIL-C-915 as applicable.

For IEC cable requirements are in IEC standards such as 60092-350 series.

8.1 Shipboard Cable for VFD Application

The adjustable speed drive (ASD) VFD cable should be able to withstand the operating conditions like repetitive 1,600 volt peak voltage spikes from low voltage (450 V) IGBT drives and at the same time not deteriorating the performance of other drive system components.

Peak voltages on a 460 V system can reach 1200 V to 1600 V, which may cause cable insulation failure, and rapid breakdown of motor insulation.

Refer to NEC-2012 110.27, IEEE-1580, and IEEE-45.8.

The new wording proposed is: "Cable installation should be designed to limit the shield voltage to less than 50 VAC at the maximum ampacity for the specific installation. If the voltage is greater than or equal to 50 VAC, then exposed shields shall be considered as live parts (in accordance with NEC-2012 110.27) and will require guarding or insulating for personnel protection and the cable manufacturer shall be consulted to determine the suitability of the cable construction to sustain excessive shield voltages."

VFD Challenges for Shipboard Electrical Power System Design, First Edition.
Mohammed M. Islam.
© 2020 by The Institute of Electrical and Electronics Engineers, Inc.
Published 2020 by John Wiley & Sons, Inc.

8.2   Cable Shielding Guide per IEEE-1143

(Extract from IEEE STD 1143-1994) IEEE Guide on Shielding Practice for Low Voltage Cables

Overall shield (Section 6.6.5)

An overall shield should be chosen for power surges, or other transients and for protection against magnetic field (inductive) coupling

Grounding and installation (Section 7)

Introduction (Section 7.1)- IEEE-1143

Power grounding is for the purposes of electrical safety and to enhance the reliability and operation of electrically operated or supplied equipment within a facility. Signal grounding, on the other hand, is to assure "noise free" operation and reliability of the electronics system. Of greater importance, however, is the harmonizing of the safety and high-frequency grounding techniques so that electrical safety is not sacrificed in order to obtain satisfactory operation of the electronic system. The National Electrical Code (NEC) (ANSI/NFPA 70-1993) should be satisfied first; then, the signal grounding should be satisfied in a compatible way that does not undo the safety aspects of the installation. It is recommended that the system safety requirements be fully understood by all concerned with the design, and then the electronic requirements be overlaid onto the electrical safety basics, in such a way as to not diminish safety.

Shield grounded at one end (Section 7.2)-IEEE-1143

This is a technique to handle low-frequency noise and may not be appropriate for high-frequency noise or noise due to transients. The shielding may be grounded at either the sending end or the receiving end.

Shield grounded at sending end (Section 7.2.1)-IEEE-1143

Grounding the overall shield of a cable or a shielded pair at the sending end eliminates the transient voltage on the cable due to the electric field. However, a different condition prevails for the magnetically induced potential. Grounding the shield at the sending end has no effect on the magnetically induced component.

Shield grounded at receiving end (Section 7.2.2)

Grounding the overall shield or a shielded pair at the receiving end again prevents the electric field from reaching the cable, eliminating the electric field component. For the magnetically induced component, the input capacitance is now the only circuit element between the voltage source and the cable. The capacitance between the cable and ground and the input capacitance forms a voltage divider. This arrangement reduces the surge voltage on the cable. The amount of reduction increases with cable length [B27].

Shield grounded at both ends (Section 7.3)-IEEE-1143

Here, for electric-field-induction, the displacement currents to the cable through capacitive coupling to the interference source are diverted to ground at each end and no transient voltage appears on the cable.

Grounding the shield at both ends completes a closed loop through the cable and ground mat system if the equipment or electronics at each grounded end are independently grounded. The magnetic field linking this loop induces a potential which in turn causes a secondary current to flow in the loop. The magnetic field due to this induced current opposes the primary field so that the net field in the loop is just sufficient to induce a potential drop related to the total resistive and reactive impedance around the current loop. This current flows axially along the shield of the cable.

Therefore, there is only a small magnetically induced voltage between the cable and the shield at the receiving end. With the shield grounded at two relatively greater in separation points, there is a risk that potential gradients in the ground mat system during faults may cause relatively larger shield currents to flow. Damage to the shielded cable may result if the shield is not a robust conductor. Therefore a heavier (thicker) shield (rather than a foil shield) is required. Use of an overall shield of corrugated 0.125 mm (0.005 in) copper or 0.2 mm (0.008 in) aluminum is usually sufficient to handle these surge currents [B4].

Transient protection with overall shields (Section 7.4)-IEEE-1143

When two electrical or electronic units are interconnected by a cable with the shield grounded at only one end, the load and source have an impedance to the common reference (usually ground). This establishes a "ground loop" between source and load, via the conductors, and the common "ground" reference. The common reference could be earth, and the example might represent wiring between two buildings. If a transient such as lightning causes fast changing electromagnetic lines of force to intercept the cable, then a current will be induced or coupled into all metallic portions of the cable except those open (ungrounded) at both ends, including the conductors and the shield(s).

The current induced into the inner conductors of the cable will be proportional to the loop impedance of each conductor, and it will be in the common-mode. This makes it difficult, for example, to detect and measure small differential currents with instrumentation connected line-to-line. In practice, the induced currents may be identical if they are common mode and very close in magnitude to one another because of similarities in loop impedances. The induced current is now capable of driving large surge currents into the source or load with the potential for destroying electronics connected to the ends of the conductors.

A cable having an overall shield of sufficient thickness can be used to compensate for the induced currents in the conductors. For a cable with a shield that is effectively grounded at both ends, there is protection from the electromagnetic influence. (NOTE—Sometimes "bonded" is used to describe the grounding of the shield. Usually, the shield represents a lower total impedance throughout its length, and its impedance to the common grounding medium at either end is lower than that of the inner conductors. The shield must be grounded continuously over its length and at both ends so that there are no random paths established through the electronic components. With a lower loop impedance, the shield can develop a greater induced "noise" or surge current than the *inner* conductors. This condition is an advantage, because as the shield carries more current, it also becomes another electromagnetic influence affecting the inner conductors. The result is a second induced current in the conductors from the current flow on shield.

However, the induced current from the shield is opposite in phase to that induced by the original transient influence. Under these conditions, the inner conductors are confronted with two induced currents, but of opposite polarity. The result is a current in the conductor which is 180° out of phase with the noise current. As a result, the induced current from the shield cancels ("bucks out") the original surge ("noise") current in the conductors.

The transient magnetic fields from the interference current (Im) induce a potential which causes transient current to flow in the shield (Is) and conductors (Ic). Since the source of the shield current (Is) and conductor current (Ic) was the same (i.e. the interference current), these currents are in phase. The shield and conductors are tightly coupled to each other in a parallel path over the entire length of cable. Therefore, the current flowing in the overall shield induces a second current (Isc) into the conductors. Isc is 180° out of phase with the shield current (Is). As a consequence, this counter current nearly cancels out the interference current in the conductors. In order to handle the magnitude of current required to effectively cancel out the interference current, the cable needs an overall shield of 0.19 mm (7.5 mil) aluminum, 0.125 mm (5 mil) copper, or other overall shield or armor material. The canceling or bucking action of the counter current is never exact, but does result in a definite and significant reduction in net induced current. This discussion has assumed that the voltage reference of the control circuit is an external ground. If the control circuit instrument is referenced to the shield, grounding the shield at both ends can increase the interference current in the conductors.

The effect of an overall shield consisting of longitudinally formed plastic coated 0.19 mm (7.5 mil) aluminum on reducing the effect of transients was verified experimentally. A test circuit of instrument cable with two pairs of twisted conductors was pulsed with a 62.5 kHz, 50 kV, 8730 A oscillatory surge. The voltages measured to ground of the overall shield and the conductors for an open or grounded shield are listed in table 7. This data shows that grounding the shield at both ends substantially reduces the induced voltage on the conductors. If the overall shield is left floating, or grounded only at one end, it will not be a complete circuit like the conductors. Therefore, counter current to cancel the interference current in the conductors cannot flow. As a result, the control circuit is not protected from the transient. The data also indicate that the use of a twisted pair, or foil shield, is not sufficient to shield signal conductors from electrical transient.

To maximize the flow of this counter current, everything that affects current flow should receive close attention. In particular, electrical continuity of the shield, electrically stable connections of low resistance, and grounds with low resistance are necessary. For two separated areas sharing a common grounding path, it is possible to drive a current directly through this ground path. Under this condition, the shield is still lower in impedance and forms a path in parallel with the signal conductors that usually have a higher impedance. Therefore, most of the noise current flows in the shield circuit rather than in the signal conductor.

Leaving the shield ungrounded ("floating the shield") will negate the ability of the shielding system to cancel electromagnetically induced currents. Overall shields grounded/bonded at both ends still retain their electrostatic shielding ability as well as provide electromagnetic shielding.

Grounding of cable with foil shields and overall shields (Section 7.5)-IEEE-1143

Grounding system (Section 7.5.1)

Grounding is applied first and foremost as a personnel safety protection device; second, as a means of limiting damage to equipment and cables; and third, for selective electrical system coordination (protection). The proper practice for modern plants is to provide single point grounding. A list of separate systems that may need to be connected to this ground is as follows:

a) Power system ground (system ground)
b) Instrument signal ground system
c) Computer signal ground system
d) Lightning protection ground system

Anywhere from one to four separate ground systems may be specified or designed. These systems use conductors to connect to each other and are grounded at one point to form a single point ground system.

The purpose of earth grounding is protection from lightning and transmission line ground faults, utilizing the earth as part of the return path for these currents. The grounding principle applicable for utilization voltages of < 600 V is not to use the "earth" as the current return path, but rather to use a system of interconnected conductors to equalize the voltage differences. This system can more effectively limit voltage differences than multiple ground connections.

For the transient and high-frequency interference control, the grounding system must be viewed as a possible interference distribution system. Transients injected onto the grounding system by power switching or other sources can propagate to all parts of the grounding conductors, including small-signal electronic circuits, that share the common grounding system. This can happen because the ground electrode has a surge impedance much greater than zero and the grounding conductors connected to it have surge impedances much less than infinity.

To prevent the grounding system from becoming an interference distribution system, the grounding must be coordinated with the shielding. That is, the local shield (equipment case, rack, shield room, or building) should serve as the grounding point for all circuits inside the shield, and grounding conductors should not penetrate any shield. This shielding and grounding topology is illustrated in Figure 19, where the shield surfaces are shown as dashed lines and the grounding conductors are shown as solid lines. The shields are closed surfaces, and the grounding conductors are not allowed to penetrate the shield surfaces. The external ground is connected to the outside surface and the internal ground is connected to the inside surface. Thus, no transient or broadband interference waves are allowed to enter the protected space on the grounding conductor.

In particular, large transients generated outside the building (such as power switching transients or lightning) can be almost totally excluded from the sensitive integrated circuits inside the equipment case, since these waves are interrupted by several layers of shielding. Yet the safety requirements are satisfied, since within any shielded volume, all cases, racks, and equipment are grounded to the local shield structure. In addition, DC and 60 Hz fault currents can flow to actuate fuses and circuit breakers, since the shields are transparent to power frequencies.

The cable shield is a part of the shield system, and it is not a grounding conductor. For a subsystem, the shield may consist of two equipment cases interconnected by a shielded cable. The complete shield (the two equipment shields and the cable shield) should be closed by connecting the cable shield to the equipment cases at each end—preferably with a circumferential bond between the cable shield and the equipment shield. The pigtail connection is adequate only for audio frequencies; at higher frequencies its inductive reactance is sufficient to produce a significant discontinuity between the shielded parts. The electrostatic shield is also a special case that cannot be applied to transient and broadband shielding.

Termination practice for overall shield (Section 7.5.3.1)-IEEE-1143

Shield bonding connectors are electrical shield terminating devices that serve a very important function in the protection of instrument and control systems from electromagnetic interference. The primary function of the connectors is to pass shield currents associated with electrical interference from the cable shield to the equipment shield without allowing the electrical interference to enter the equipment. Therefore, electrical stability of the connectors is essential.

An evolution in the design of connectors has resulted in connectors suitable for cable with plastic coated or bare overall shields. These connectors do not require the jacket to be stripped when the jacket is bonded adhesively to the overall shield. The connectors are slipped, over both the jacket and overall shield, using small tangs to lock into the jacket and to penetrate the plastic coatings on the inner side of the shield. Sophisticated test requirements have been developed by the telecommunications industry to assure reliable performance of the connectors. Generally, the following types of tests should be passed successfully for acceptance:

a) Connector resistance
b) Environmental requirements—vibration, temperature cycling, hydrogen sulfide exposure, and salt fog exposure
c) Endurance tests—fault current and current surge

A common treatment of a shield at a connector is to insulate the shield with tape and connect it to the back shell of the equipment housing or case through a pigtail. The tail can introduce into the circuit an inductance of about 1000 nH/m, an inductance much higher than the transfer inductance produced by an equal length of overall shield. The shield current flowing through the inductance of the pigtail creates an interference voltage between the cable shield and the equipment case. It is important to keep grounding connections to the shield as short as possible.

For certain cables it may be possible to terminate the overall shield concentrically on the connector shell with no gaps in the circumference. When this procedure is followed, there is much less voltage introduced into the conductors for the following reasons :

a) Length of the path through which the shield current must flow is shorter.
b) Field intensity inside the shield is nearly zero because the magnetic field is entirely external to the region occupied by the signal conductors (major factor).
c) Field intensity external to the shield is reduced by virtue of the inherently larger diameter of the path upon which the current flows (minor factor).

8.3 VFD Cable Characteristics and Termination Techniques

See Figures 8.1 through 8.6.

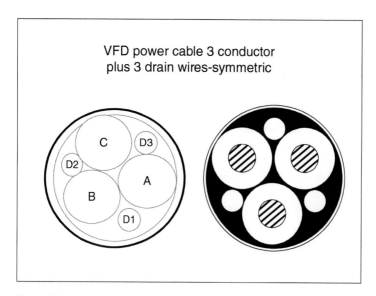

Figure 8.1

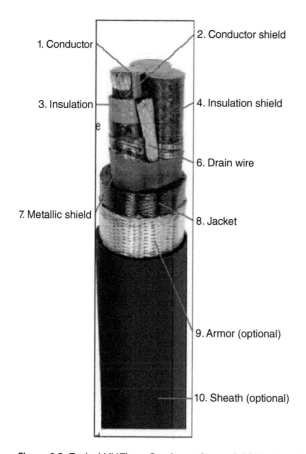

Figure 8.2 Typical MV Three Conductor Power Cable (Adapted).

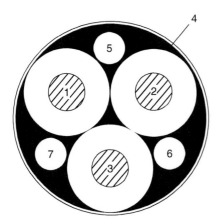

- Item 1: Basic cable with conductor and insulation
- Item 2: Basic cable with conductor and insulation
- Item 3: Basic cable with conductor and insulation
- Item 4: Outer overall insulation
- Item 5: Interstice ground conductor - bare
- Item 6: Interstice ground conductor - bare
- Item 7: Interstice ground conductor - bare

Figure 8.3 Typical Power System MV Three Conductor VFD Cable Details - Without Armor (Adapted).

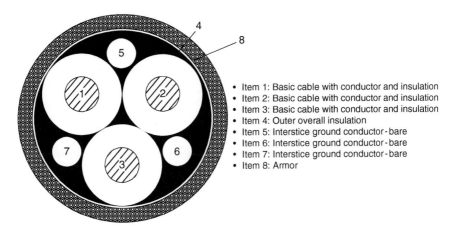

- Item 1: Basic cable with conductor and insulation
- Item 2: Basic cable with conductor and insulation
- Item 3: Basic cable with conductor and insulation
- Item 4: Outer overall insulation
- Item 5: Interstice ground conductor - bare
- Item 6: Interstice ground conductor - bare
- Item 7: Interstice ground conductor - bare
- Item 8: Armor

Figure 8.4 Typical Power System MV Single Conductor VFD Cable Details With Armor (Adapted).

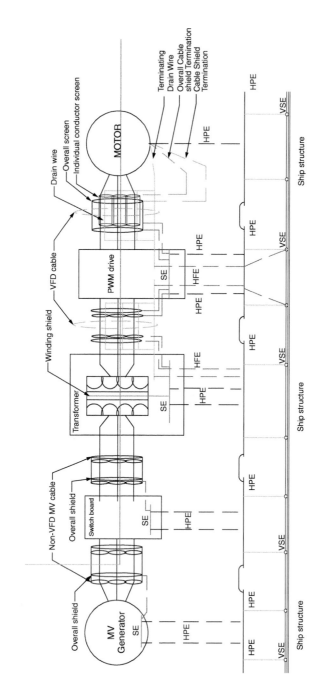

Figure 8.5 Typical VFD Cable Schematics Showing Generator to Motor (Sample-1).

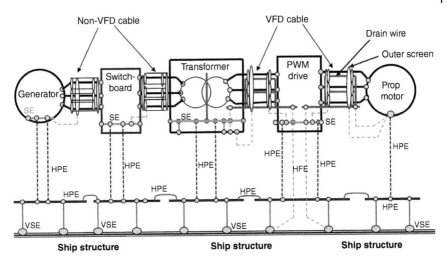

Figure 8.6 Typical Power System Cable (Sample 2).

8.4 VFD Cable Issues for Shipboard Application

The VFD cable should be able to withstand operating conditions like repetitive 1,600 volt peak voltage spikes from low voltage IGBT drives and at the same time not deteriorate the performance of other drive system components. Peak voltages on a 460 V system can reach 1200 V to 1600 V, causing rapid breakdown of motor insulation, leading to motor failure. If this is left uncontrolled, insulation failure may occur.

The same peak voltages that damage the motor can also damage the cable. In the perfect cable power delivery system the net instantaneous current flowing in the total cable system should be zero. This includes all phase conductors, all ground conductors and shield. This can be achieved by a symmetrical cable.

8.4.1 Symmetrical Cable

In a VFD installation the IGBT switches are in constant operation at high frequency and this produces an inverter output voltage with a PWM wave. This IGBT switching also causes a motor line to ground voltage, normally called a common mode voltage. Most AC drives, in addition to their normal three-phase output voltages, create a fourth unintended voltage to ground, known as common mode voltage.

Common mode current is current that leaves a source and does not come back to the source. In most closed loop electrical circuits, most of the current returns to the source. However, there is a small amount of current that in any circuit is radiated and does not return.

The common mode voltages cause short high-frequency pulses of common mode current to flow in the safety earth circuits, and it is essential that the common mode currents return to the inverter without causing EMC-EMI problems in other equipment, and this means that the common mode currents must not flow in the safety earthing (grounding) system.

The best and easiest way to do this is to use shielded VFD cables that are properly terminated and provide a low impedance path for common mode current to return to the inverter.

8.5 ABS Steel Vessel Rule: Part 4 Chapter 8 Section 4 – Shipboard Cable Application

8.5.1 Single Conductor Cables

As far as possible, twin-or multi-conductor cables are to be used in AC power distribution systems. However, where it is necessary to use single-conductor cables in circuits rated more than 20 A, arrangements are to be made to account for the harmful effect of electromagnetic induction as follows:

i) The cable is to be supported on non-fragile insulators.
ii) The cable armoring (to be non-magnetic, see 4-8-3/9.11) or any metallic protection (non-magnetic) is to be earthed at mid span or supply end only.
iii) There are to be no magnetic circuits around individual cables and no magnetic materials between cables installed as a group.
iv) As far as practicable, cables for three-phase distribution are to be installed in groups, each group is to comprise cables of the three-phases (360 electrical degrees). Cables with runs of 30 m (100 ft) or longer and having cross-sectional area of 185 mm^2 (365,005 circ. mils) or more are to be transposed throughout the length at intervals of not exceeding 15 m (50 ft) in order to equalize to some degree the impedance of the three-phase circuits. Alternatively, such cables are to be installed in trefoil formation.

8.5.2 VFD Cables Selection Criteria

Cable used to connect the VFD drive to the motor affect how well the VFD system handles the phenomena with potential to disrupt operations. Cables designed specifically for VFD applications (see Figure 1) withstand these phenomena. And, equally important, they do nothing to make the situation worse. A generic motor supply cable which is not designed with the specific requirements of VFD system in mind can actually cause problems.

Here are some of the things to look for in a VFD cable.

8.5.2.1 Insulation

The insulation must have excellent dielectric properties to prevent breakdown from the stresses of voltage and current spike, corona discharges, and so forth.

Cross-linked polyethylene (XLPE) has the required properties, making it a better choice than standard polyethylene or PVC.

XLPE insulation helps reduce the potential effects of harmonics and corona discharge, therefore providing stable electrical performance, which prolongs the life of the cable. This reduces the need for costly downtime due to cable failure.

8.5.2.2 Shielding

Shielding serves the double purpose of keeping noise generated by the VFD cable from escaping and preventing noise generated outside the system from being picked up. There are three types of shields typically used:

1) Foil Shield
   Foil shielding uses a thin layer of aluminum, typically laminated to a substrate such as polyester to add strength and ruggedness. It provides 100% coverage of the conductors it surrounds, which is good. It is thin, which helps keep cable diameters small, but is it harder to work with, especially when applying a connector.
2) A woven braid
   A woven braid provides a low-resistance path to ground and is much easier to terminate by crimping or soldering when attaching a connector. Depending on the tightness of the weave braids typically provide between 70% and 95% coverage. Because copper has higher conductivity than aluminum and the braid has more bulk for conducting noise, the braid is more effective as a shield. But it adds size and cost to the cable.
3) Both foil and braid shields
   A third approach combines both foil and braid shields in protecting the cable. Each supports the other, overcoming the limitations of one with its own compensating strengths. As shown in Figure 8.2, this presents shielding effectiveness superior to either approach alone.
   One purpose of the shield is to provide a low-impedance path to ground. Any noise on the shield thus passes to ground. A poor ground connection can increase the impedance, which also increases the potential for noise to be coupled to nearby cables or equipment.

8.5.2.3 Cable Geometry

A round, symmetrical cable gives the best electrical performance and resistance to deleterious effects. One reason is that round symmetry creates more uniform electrical characteristics in the cable. The electrical characteristic of one conductor does not differ significantly from another. For example, having the ground wires places symmetrically around the cable provides a more balanced grounding system.

8.5.3 The Right Cable Prevents Problems

The cable selected for interconnecting a variable frequency drive to a motor can significantly influence the reliability of the system.

The wrong cable can increase problems and lead to premature failure.

The right cable, on the other hand, can actually reduce the potential from problems like corona discharge or standing waves.

8.6 Drain Wire

The effect of an out of balance three-phase load causing the neutral conductor to become a heat emitter should not cause any overheating to the group of three-phase four-wire cable. However, when harmonics are present as shown in Figure 1(b), the line current in the three phases are no longer balanced sine waves and if they are in triplen order (i.e. 3n), they will be additive in the neutral. Now the neutral conductor becomes a fourth and additional current carrying conductor. As a result, it is an additional heat-emitting source in the group of four conductors.

In view of the fact that the neutral conductor is now a current carrying conductor, hence a heat emitter, steps need to be taken to take account of the extra heat that is produced by the neutral conductor in a three-phase circuit. The update published by the IEE (now the IET) [13] states that for every 8 °C increase above the maximum core conductor continuous operating temperature the life of the cable will be halved (e.g. 25 years reduced to 12.5 years). A method is thus required to size the cable accordingly to dissipate the extra heat that is being generated within a group of three-phase four-wire conductors to ensure that the group of four conductors does not.

8.6.1 Using Separate Neutral Conductors

On three-phase branch circuits, another philosophy is to not combine neutrals, but to run separate neutral conductors for each phase conductor. This increases the copper use by 33%. While this successfully eliminates the addition of the harmonic currents on the branch circuit neutrals, the panel board neutral bus and feeder neutral conductor still must be oversized.

8.7   IEEE-45.8 VFD Cable Application Extract

8.7.1   Variable Frequency Drive Applications
(Section 5.3 of IEEE-45.8)

Cables for VFD applications require special consideration because of harmonics, electromagnetic interference (EMI), reflected wave voltages, common mode current, and induced voltages in adjacent cables.

The interconnecting cable between the motor and drive is an integral part of the entire system. Every variable frequency drive system is different and, typically has manufacturer's recommendations for cables and installation. Therefore, consultation between the system designer, drive manufacturer, and the cable manufacturer is recommended for proper cable selection.

In addition to the guidelines that follow, cables used in variable frequency drive (VFD) power applications should meet all of the requirements of Table 6 in IEEE-1580.

8.7.2   Construction Notes (Section 5.3.1 of IEEE-45.8)

8.7.2.1   Insulation (Section 5.3.1.1 of IEEE-45.8)

For selecting VFD motor controller power supply cable, depending on the nominal voltage rating and the reflected voltage, over insulated conductor ratings should be considered (Example: If the VFD has a nominal voltage rating of 600 V, there is the potential for reflected voltages of up to 3X this voltage. Thus, the cable should have a 2 kV voltage rating to reduce the potential for early failure.). Selection of the insulation thickness will also be dependent on the switching frequency and length of cable installed. Reducing the capacitance of the cable reduces common mode current. It is recommended that the insulation material used on these cables have a dielectric constant less than 3.0 to reduce cable capacitance. PVC should not be used as an insulating material.

8.8   VFD System Stray Noise Circulation Details

See Figure 8.7.

8.9   VFD Cable Installation Recommended

The issues of VFD cable are VFD-generated common mode currents (CMC) and electromagnetic radiation. Both are generated as an unwanted byproduct of the drive's high frequency pulse width modulated waveforms. The cable

VFD motor controller proper cable selection, cable routing and cable termination for power cable, control cable, signal cable

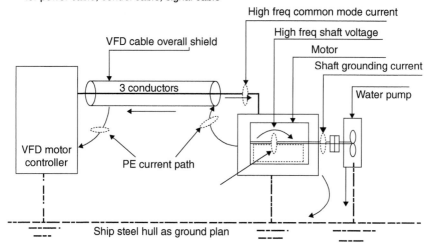

Figure 8.7 VFD System Stray Noise Circulation Details.

termination must be as recommended by the VFD manufacturer with the following considerations:

1) Provide a managed and controlled path for the VFD generated common mode current to ensure minimum CMC current travels through the ship's steel hull.
2) Provide a continuous shield over the entire cable from the inverter to the motor (minimizing the amount of electromagnetic radiation that can escape from the cable which can cause interference with other systems).
3) Effectively ground the VFD motor frame (reducing shock hazard).

The drive manufacturers recommend that a VFD cable shield be bonded at both ends to provide an effective path for the common mode current. The cable termination should also be in electrical contact with the cable shield on all sides (360°) to provide the low impedance path from the cable shield to the entry point at the VFD motor controller enclosure and the motor. This termination method also minimizes electromagnetic radiation from the cable.

In addition to proper terminations, make sure the cable jacket is left intact between the inverter and the motor terminations. The cable jacket will act as an insulator to prevent common mode current traveling down the cable's shield from jumping off to the building ground.

It is a wise practice to keep input/output cables to one side of a cabinet and separate any Programmable Logic Controller (PLC) and other control and communication equipment cables to the opposite side of the cabinet to reduce the effects of electromagnetic interference from the drives.

Marking the Jacket...

A) Insert the cable into the enclosure. Securely route the cable through the enclosure to the drive (or motor) terminals allowing a sufficient length of cable to be routed to minimize cable strain.
B) Measure the length of cable necessary to reach the designated termination points, allowing excess length for trimming once the cable is routed to the termination points after preparation.
C) Mark the jacket at the enclosure entry point. This is where the cable jacket will be cut and stripped.

Removing the Jacket...

A) Wrap electrical tape around the jacket at the mark previously made with the leading edge of the tape lined up with the mark.
B) Cut around the jacket along the leading edge of the tape down to the shield layer without cutting through the shield.
C) Cut longitudinally from the cable end to the leading edge of the tape down to the shield layer.
D) Remove the jacket.

Removing the Shield...

A) Measure two inches from the previously cut edge of the jacket and mark.
B) Wrap a strip of copper tape (length equal to approximately one and a half times the circumference of the shield) around the shield with the leading edge of the tape lined up with the mark.
C) Align and wrap electrical tape over the copper tape (for contrast with shield when cutting), and then cut around the shield at the leading edge of the tape down to the cabled assembly layer.
D) Remove the shield and remove the electrical tape applied in "C" leaving the copper tape.

Attaching the Ground Strap...

A) Remove fillers. Twist drain wires together and wrap wires with electrical tape.
B) Place soldered section of ground strap flush against the shield. Wrap ground strap around shield while maintaining tension to avoid spreading the strap's braid.
C) Secure ground strap to shield by applying a constant force spring in the same direction as the shield was wrapped.
D) Wrap electrical tape around the spring and ground strap to secure to the shield.

Terminating the Ground Strap...

A) If you will be securing the cable to the enclosure (or motor junction box) with a cable fitting (gland), place the fitting (gland) over the cable before proceeding.
B) Broaden the exposed end of the ground strap.
C) Clean the back plate surface so when the ground strap is connected to it, there is a good low impedance path for the common mode current flowing through the ground strap.
D) Attach the ground strap and wrapped drain wire(s) (if present) to the back plate surface using a large flat washer. If installing the termination at the motor end of the cable, attach the ground strap and wrapped drain wire(s) (if present) to a wall of the junction box.

Terminating the Conductors...

A) Route the phase conductors and grounds to the proper termination points in both the drive enclosure and the motor junction box.
B) Trim away any excess length while leaving sufficient material for phase reversal if required.
C) Attach lugs to the phase conductors in accordance with the drive and motor manufacturers' instructions

8.10 Shipboard Grounding and Bonding Termination Recommendation

The VFD cabinet ground bus shall be connected to the hull at point and the point must be accessible for taking measurements. If multiple VFD drive modules are in one cabinet enclosure, all VFD modules must be grounded for personal safety reasons to prevent unwanted voltage build-up under any circumstances.

The ground connection though cabinet fixing screws and cabinet chassis is not recommended. To ensure proper ground-bonding connection the module ground must be connected through approved method. The high frequency and low impedance grounding connection should be around 0.1 ohm. This can be achieved by using with a flat cooper braid bonding as shown in Figure 8.8.

Use a ground strap instead of solid round wire for grounding. That addresses the skin effect so prominent in high-frequency VFD signals.

Because VFDs have unbalanced three-phase vector sums, current is never equal to zero at neutral. That necessitates proper shielding and grounding paths with low impedance back to the drive (and to protect against the effects of common-mode currents and high-frequency bearing currents).

Figure 8.8 Recommended Grounding Connectors.

Never use unshielded cables in conduit or tray to prevent unintended paths to earth ground, creating ground loops. In addition, cables must have proper shielding and grounding where they connect to the drive enclosure and where they connect to the motor. For this, use EMC glands and shielded cables.

8.11 EMC Cable Glands

EMC cable glands provide a 360° contact with the enclosure for better paths to ground. In fact, grounding EMC glands can be used on both the motor and enclosure. Some are nickel-plated brass and rated IP 68; EMC glands are available in NPT and metric threads and sizes. In addition to the 360° termination at the enclosure, the shield's pigtail should be kept as short as possible when tied to the PE terminal at the drive.

8.11.1 Cable Length Between VFD Motor Controller and Motor

Let us start with long cable runs between the variable frequency drive motor controller and motor with regards to proper VFD operation:

1) The capacitance from the cables (phase to ground) can create enough current to ground where there can be nuisance malfunction of digital signals and equipment operation.
2) These limits are only around ground currents (where the VFD would not properly operate). There are other considerations such as damaging motor bearing, motor insulation stress and VFD cable insulation stress.

3) For motor insulation protection, there are several considerations. Specify motor insulation to be at least 1400 V.
4) Specify VFD cable to 2 kV.
5) There are recommendations for using different types of filters for various VFD applications. The precaution is that the use of any filter related to the EMI/EMC management must be considered as a system issue not only as an individual VFD issue.

Due to the skin effect at high frequencies, do not use a single terminated drain wire. It does not provide adequate surface area to conduct the high frequency noise to ground. Rather, it is best to tie the shield directly to the PE terminal at the drive. Because VFD signals have a very high frequency, they exhibit the skin effect on wires at those high frequencies.

As a review, the skin effect is a tendency for alternating current to flow mostly near the outer surface of an electrical conductor. The effect becomes more and more apparent as the frequency increases.

This is also why it is best to use a ground strap instead of solid round wire for grounding.

9

Ship Smart System Design (S3D) and Digital Twin

The ship smart system design (S3D) is introduced for the ship system design engineers to address the design challenges of VFD motor controller. The S3D concept was introduced by the author in 2003 (See Figure 9.1). The S3D concept is also recommended for existing a VFD and ASD-populated motor controller based electrical system to establish exact nature of VFD generated electric noise to compare with the calculated noise and then establish a baseline if it is within the regulatory requirements. If the electrical noise (harmonics) level is more than the acceptable level, then undertake Harmonic Management System to bring the harmonics to an acceptable.

The S3D concept evolving over the past two decades has been enabled by the advancement of low-cost computing, powerful software, and analytical platforms for data mining, machine learning, and data networking technology. Now S3D is embodied in "Industry 4.0-Digital twin" as being deployed for simulated, and proven ship system design, development starting from concept phase to ship delivery, operation, maintenance, and training.

A "Industry 4.0-Digital twin" is a digital representation of an object or system (e.g. a ship, an engine, individual equipment control function, or centralized control system). It can contain various digital models, data repositories, and processes related to the real object. Data can be stored in several representations, including a dynamic simulation model, graphical 3D-models, analytical models, database schemas, sensor settings, and measurements, etc. By designing physical assets in a virtual environment using digital twins, new attributes can be developed over much shorter time periods. Open source frameworks will allow designers, equipment manufacturers, ship builders, ship owners, and ship operators to efficiently collaborate on building and improving objects, drastically shortening the innovation cycle. Digital twin information can be used continuously throughout the design, development, procurement, production cycles, as well as throughout the life cycle of the ship.

VFD Challenges for Shipboard Electrical Power System Design, First Edition.
Mohammed M. Islam.
© 2020 by The Institute of Electrical and Electronics Engineers, Inc.
Published 2020 by John Wiley & Sons, Inc.

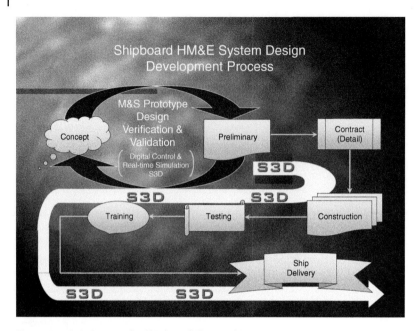

Figure 9.1 S3D Concept for Shipboard Electrical System Design and Development.

The potential benefits of getting S3D to reality using digital twin is enormous. Throughout the life cycle of a vessel, production, operation, and maintenance processes can be simulated and optimized for maximum business return on investment. Similarly, modifications to the digital twin can predict performance improvement, and can be accomplished with minimum cost burden. The subsets of knowledge can be embedded into the vessel's automation to create artificially intelligent machinery that continuously minimize operational risk and associated costs.

9.1 Smart Ship System Design (S3D) Overview - VFD and ASD

The S3D concept is shown in Figures 9.1 and 9.2 specifically addressed for shipboard electric propulsion with variable frequency drive (VFD) motor controller and adjustable speed drive (ASD) motor controller. See Table 9.1. For regular requirements such USCG regulation, IEEE-45 Recommendations and ABS rules, see Table 9.2 for system breakdown for S3D based simulation.

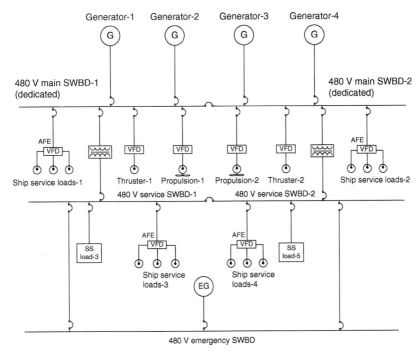

Figure 9.2 Typical Electrical One Line Diagram For Shipboard Low Voltage VFD Propulsion System and Ship Service System With VFD.

Table 9.1 Shipboard Electrical Power System Design Attributes by Regulation.

Item	Requirements	Regulations	Remarks
1	N+1 Power Source	USCG, ABS, IEEE-45	Ship service power generation
2	Redundancy Requirements	USCG, ABS, IEEE-45	Ship service power generation, and vital equipment
3	Vital And Non-Vital Classifications	USCG, ABS, IEEE-45	Survivability related equipment
4	Voltage Drop	USCG, ABS, IEEE-45	
5	Emergency Power	USCG, ABS, IEEE-45	Survivability related emergency power

(Continued)

Table 9.1 (Continued)

Item	Requirements	Regulations	Remarks
6	Emergency Loads	USCG, ABS, IEEE-45	Survivability related emergency load
7	Transients-Frequency	USCG, ABS, IEEE-45	Transient limitations
8	HARMONICS	USCG, ABS, IEEE-45	Harmonic limitations
9	UPS	USCG, ABS, IEEE-45	Uninterruptible power
10	Sensitive Loads	USCG, ABS, IEEE-45	Loads sensitive to transients, and electrical noise
11	Electrical System Protection System	USCG, ABS, IEEE-45	Ensure overload management
12	Grounding - issues	USCG, ABS, IEEE-45	Ensure ground protection-system level
13	Proper Voltage Level Selection	USCG, ABS, IEEE-45	
14	Electrical Plant Load Analysis	USCG, ABS, IEEE-45	
15	Power Generation-selection of generators	USCG, ABS, IEEE-45	
16	Electrical Power Distribution System – Ring Bus, Radial Bus, Zonal Distribution	USCG, ABS, IEEE-45	
17	Dedicated bus for VFD & ASD loads	IEEE-45	
18	System Level Protection and System Coordination	USCG, ABS, IEEE-45	
19	Survivability-System	USCG, ABS, IEEE-45	
20	VFD & ASD Motor Controller Selection	USCG, ABS, IEEE-45	
21	VFD Type cable Requirements	IEEE-45	

VFD Topology–IPS design and THD boundary. The propulsion SWBD THD level could be higher than 5% because this is the way the system is designed. However, the ship service SWBD must be max 5% one section of the ship service SWBD is provided with THD less Than 5% to meet the system requirement. The THD level must be simulated. Calculated and then measured to establish the real THD at the delivery of the vessel.

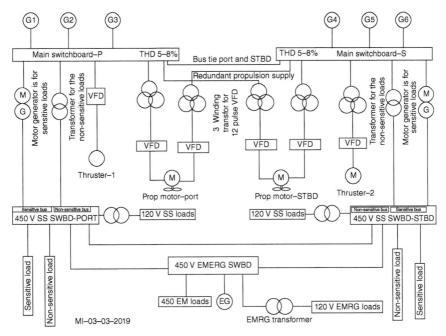

Figure 9.3 Typical Electrical One Line Diagram for Shipboard Medium Voltage VFD Propulsion System.

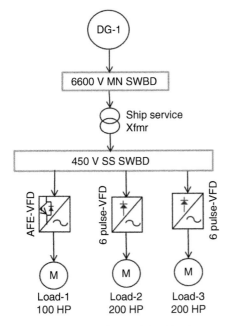

Figure 9.4 Typical Electrical One Line Diagram for Shipboard Auxiliaries With PWM AFE and 6 Pulse Rectifiers.

Table 9.2 Shipboard Electrical Power System Simulation Based Design With S3D.

Item	Requirements	S3D Attributes	Remarks
1	N+1 Power Source	Simulation based hardware in the loop starting from concept	
2	Redundancy Requirements		
3	Vital And Non-Vital Classifications		
4	Voltage Drop		
5	Emergency Power		
6	Emergency Loads		
7	Transients-Frequency		
8	Harmonics		
9	UPS		
10	Sensitive Loads		
11	Electrical System Protection System		
12	Grounding		
13	Proper Voltage Level Selection		
14	Electrical Plant Load Analysis		
15	Power Generation-selection of generators		
16	Electrical Power Distribution System – Ring Bus, Radial Bus, Zonal Distribution, etc		
17	System Level Protection and System Coordination		
18	Survivability-System		
19	Propulsion VFD Selection		

9.2 Ship Design Regulatory Fundamentals – Electrical System Design and Development

These attributes are fundamentals to ensure safety for the ship as well as shipboard personnel. See Table 9.3.

Table 9.3 S3D Based Modelling and Simulation Fundamentals.

Attribute	Concept Design	Preliminary Design	Detail Design	Remarks
1) Proper Voltage Level Selection				
2) Propulsion VFD				
3) System total harmonic distortion				
4) Electrical Load Analysis				
5) Electrical System – Ring Bus, Radial Bus, Zonal Distribution, etc.				
6) Grounding – System Level – Continuous monitoring				
7) System Level Protection and System Coordination				
8) Performance analysis using physics Model				
9) System Dynamic Analysis				
10) Power Flow Management				
11) Survivability-System				
12) Graceful System Degradation				
13) Automatic Fault Management – System Level				
14) Redundancy compliance				
15) Standards, Rules and Regulations				

9.3 S3D Based Modelling and Simulation for Design Verification

See Table 9.4.

9.4 Sample System Level THD Requirements

Requirements are as follows:

a) **6600 V Propulsion Motor Adjustable Speed Drive (ASD) Bus-Harmonic Distortion Factor**

Limit the voltage total harmonic distortion (THD) on the high voltage common bus, as follows:

  i) THD no more than 8% (assumed), at full motor speed, and under all normal steady state operating conditions within the motor duty cycle, with both twelve-pulse drives operating on-line and with all four main generators connected to the central power plant high voltage bus.

  ii) THD no more than 15% (assumed) under transient duty conditions of maneuvering and/or propeller max interaction with both twelve-pulse drives operating on-line and with all four main generators connected to the central power plant high voltage bus. Transient duty shall assume both motors operating at 175% of rated torque and zero speed for 30 seconds. (assumed)

  iii) THD no more than 12% (assumed) under all steady state operating conditions with one converter operating in a six-pulse mode (maximum output at half power), and with the other converter system operating normally in a twelve-pulse mode. This is a condition where one propulsion motor is operated at full power and the other propulsion motor is operated at 50% power with all four main generators connected to the central power plant high voltage bus.

  The harmonic analysis report shall include expected harmonics caused by the loading of the converter. The report shall identify system ratings and analytical assumptions, identify frequency of converter input harmonics considered, tabulate the magnitude of input current harmonics and present these values in a graphical representation of the waveform, and tabulate magnitudes of the voltage harmonics for the motor speeds and motor currents:

  These tabulations of voltage harmonics shall be provided for viable combinations of numbers of main generators and main motors on-line at a given time. The report shall provide recommended ratings, based on the current and voltage harmonics, for the major components of the central power plant and the electric propulsion system. These recommendations shall be used to select the equipment.

b) **Low Voltage 690 V and Below) Propulsion Motor Adjustable Speed Drive (ASD) Bus-Harmonic Distortion Factor**

Low Voltage Ship Sensitive Service Bus – THD Limits (Refer To Figure 9.2, 9.3, and 9.4). Details are shown in Table 9.4.

c) **Low Voltage Ship Non-sensitive Service Bus – THD limits (Refer to Table 9.5).**

Table 9.4

Waveform voltage distortion	
a) Maximum total harmonic distortion	5%
b) Maximum single harmonic	3%
c) Maximum deviation factor	5%

Table 9.5

Waveform voltage distortion	
a) Maximum total harmonic distortion	5% to 8%

9.5 Variable Frequency Drive – Solid State Devices Carrier Frequency Ranges

Solid state device carrier frequency – (Device turn on and turn off)

1) SCRs (Silicon controlled rectifier) – What is the rise time of the variable speed drive's output IGBTs
    a) Can operate in the 250 to 500 Hz range, (4 to 8 times fundamentals)
2) BJTs (Bipolar Junction Transistor) can operate in the 1 to 2 kHz range (16 to 32 times fundamentals)
3) Insulated gate bipolar transistors (IGBT) for the inverter section. IGBTs can turn on and off at a much higher frequency, 2 kHz. to 20 kHz. (30 to 160 times fundamentals)
4) Typical AFE (Active front end with IGBT) switching frequency 3600 Hz (60 times fundamentals)

Refer to Table 9.6 on next page.

The EMI is also produced by the harmonics which are generated by the carrier frequencies, and rise times. The reflected waves are caused by the capacitive effects of motor leads and the resulting impedance mis-match between the motor cables and the motor windings.

The EMI and RFI travel to the motor along the motor leads and transmitted to ground via the capacitive effect between the motor windings and the motor frame, the capacitive effect between the line conductors and bond wire, and the capacitive effect between the line conductors. The EMI and RFI will then seek a return path to the source, and input of the VFD.

Table 9.6

Device Type	Carrier Frequency	Explanation	Remarks
SCR	250 Hz to 500 Hz	4 to 8 time the fundamentals	• A variable frequency drive with SCR will generate harmonics
BJT	1 kHz to 2 kHz	8 to 16 times the fundamentals	• A variable frequency drive with BJT will generate harmonics. • A variable frequency drive with BJT will generate radio frequency interference (RFI) in the range of 0.5 MHz to 1.7 MHz, and electromagnetic interference frequencies (EMI) in the range of 1.7 MHz to 30 MHz.
IGBT	2 kHz to 20 kHz	16 to 160 times the fundamentals	• A variable frequency drive with IGBT will generate harmonics. • A variable frequency drive (VFD) with IGBT will generate radio frequency interference (RFI) in the range of 0.5 MHz to 1.7 MHz, and electromagnetic interference frequencies (EMI) in the range of 1.7 MHz to 30 MHz.

9.6 Risk Factors for VFD and ASD Application-Shipboard Power System

9.6.1 Industry Standards-IEEE-45 Requirement of Harmonic Distortion

Waveform voltage distortion – maximum total harmonic distortion is 5% to 8% and maximum single harmonic is 3%. It is to be noted that these requirements are fundamental, which means all other design requirements must be complied with.

9.6.1.1 Lack of Awareness

A key risk factor is the lack of proper documentation, awareness nd training of the shipboard power system design with VFD motor controller such as:

i) VFD motor controller will have harmonic type noise which may affect entire power system. What are the trade-offs of using VFD motor controller versus other type of motor controller.
ii) Proper understanding of VFD selection, inverter type motor selection.
iii) VFD type cable selection as well as proper termination.
iv) VFD operated motor bearing current related potential problems.
v) Proper grounding of the VFD motor controller and associated equipment.

9.6.1.2 Improper/Inadequate Grounding

Improper grounding interferes with the normal flow of current through the motor frame and the grounding conductors, and fails to provide a low-impedance path. For motors on VFD motor controller, the common-mode voltage makes proper grounding even more crucial. Because the common-mode voltage ensures that there will always be more substantial ground current, it may be difficult to prevent bearing damage and system level equipment mal-functions.

9.6.1.3 Proper VFD Cable With Drain Wires And Shielding

The VFD generated voltage and current imbalance should be managed with VFD cable with proper insulation rating, drain wires, and shielding.

9.6.1.4 High Carrier Frequencies

VFDs operated at high carrier frequencies lead to a higher rate of current discharge pulses. Operating motors at higher carrier frequency may reduce motor ambient noise but the risk of motor damage increases.

Appendices

Appendix A   Equations

Generator full load current (in Amps)
$$= (\text{given KW rating}) \frac{Generator\ KW}{Voltage \times \sqrt{3}\ x\ PF} \tag{1}$$

Short Circuit current (in Amps), (given sub-transient reactance)
$$= \frac{Generator\ Full\ Load\ Current}{Generator\ Sub-Transient\ Reactancex} \tag{2}$$

$$\text{Short circuit MVA} = \left(\sqrt{3} \times \text{Voltage} \times \text{Generator Short Circuit Current} \right) \tag{3}$$

$$\text{Generator reactance (in Ohms)} = \frac{V^2}{Short\ Circuit\ MVA} \tag{4}$$

$$\textbf{THD (Total Harmonic Distortion)} = \frac{\sqrt{\sum_{h=2}^{\infty} V^2}}{V1} \tag{5}$$

V_h = RMS amplitude of voltage at the harmonic order "h"
V_1 = RMS amplitude of the 60 Hz fundamental voltage

$$\textbf{Capacitors : Capacitance}\ \ C = \epsilon \frac{A}{d} \tag{6}$$

$$\text{Ic} = C \frac{dv}{dt} \tag{7}$$

$$\text{Ic} = j2\pi f CV \tag{8}$$

VFD Challenges for Shipboard Electrical Power System Design, First Edition.
Mohammed M. Islam.
© 2020 by The Institute of Electrical and Electronics Engineers, Inc.
Published 2020 by John Wiley & Sons, Inc.

Electromagnetic or Inductive Coupling

a) The magnitude of the electromagnetic coupled voltage depends on

    i) The rate of current change $V = L\dfrac{di}{dt}$ (9)

    ii) *Separation distance of the conductor from the source of disturbance*

    iii) *Size of the signal loop area*

Appendix B   VFD to Motor Capacitive Ground Path

Variable frequency drive (VFD) motor controller for shipboard use

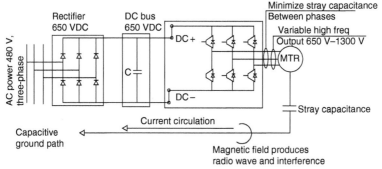

Notes :
1. Do not anchor the VFD driven motor winding neutral point to the ground.
2. If the main supply line is grounded, the capacitive ground will circulate throughout the system ground and will contaminate the entire system. Shipboard power system is usually ungrounded. The shipboard ungrounded power supply will also be effected throughout through stray ground path.
3. Minimize stray system capacitance minimize stray capacitance between cable conductors.
4. Use VFD rated cable with proper voltage rating, drain wire and shielding
5. Use short length VFD cable as recommended by the VFD supplier

Appendix C   Proper System Grounding to Manage Common Mode Current

VFD generated high frequency common mode current flows from phase to ground through distributed system capacitance. The system capacitance is between the motor stator windings and the rotor and between the motor stator windings and the motor frame. The motor rotor can become "charged" and then discharged through the motor bearings to ground or through the connected load to ground.

It is very important to provide a ground strap between the motor base and the load base to establish proper low impedance ground path.

This feature is important for both personnel safety and to prevent current flow through the bearings from the motor frame, through the bearings then to ground through the load.

Appendix D   Grounded System Balance and Unbalanced Load Calculation

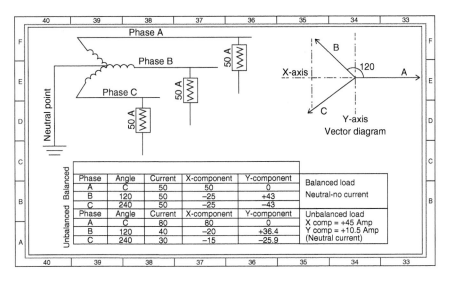

	Phase	Angle	Current	X-component	Y-component	
Balanced	A	C	50	50	0	Balanced load
	B	120	50	−25	+43	Neutral-no current
	C	240	50	−25	−43	
Unbalanced	Phase	Angle	Current	X-component	Y-component	Unbalanced load
	A	C	80	80	0	X comp = +45 Amp
	B	120	40	−20	+36.4	Y comp = +10.5 Amp
	C	240	30	−15	−25.9	(Neutral current)

Shipboard power supply is three-phase, ungrounded, sinusoidal balanced, and symmetrical under normal conditions. This means that the vector sum of the three-phases is always equal to zero.

However, the VFD fundamentals are different. The sinusoidal and symmetric power supply is converted to DC and then converted to modulated power with varying frequency at a such higher frequency then the fundamental frequency. But that's not true for VFDs variable frequency drive motor controller. The VFD inverter uses DC pulses to create three-phase voltages.

(The VFD output voltages may be symmetrical, but it is impossible to make the sum of the voltages zero.) The resulting neutral-point voltage – the center winding junction in a star-wound motor – is not zero. This unbalanced voltage is the common-mode voltage source. So, whenever the sum of the three-phase voltage levels at this junction changes, a current proportional to that change must flow and circulate in an available low resistant path. This current flow through the system components contributes to detrimental current loop, often called ground loops.

Use VFD cable with the features of an overall shield and symmetrical grounds.

Appendix E Variable Frequency Drive (VFD) High Frequency Noise Management

1 Introduction

There are major challenges as to the VFD high frequency noise identification and noise management. The low voltage auxiliary systems are being supplied with VFD motor controllers. Sometimes it is viewed as "This is a very small drive and then the VFD is the perfect application". The shipboard VFD application decision is complex with many drives with some type of ground references and steel hull boundary. Every VFD produces electrical noise which must be considered for its emi/rfi impact and then mange accordingly. IEEE-45.1-2017 has done excellent job addressing the fundamental design decision issues. The following sections are direct quote from the IEEE-45.1-2017. The IEEE-45.8 Clause 5.5 VFD application has also been quoted for ready reference.

2 IEEE Std 45.1-2017 Recommended Practice for Electrical Installations on Shipboard—Design

2.1 Clause 14. Adjustable Speed Drive (ASD) Applications States:
"The system or sub-system designer will determine when an ASD/motor combination is appropriate in consultation with the customer. The system or sub-system designer should work with the ASD supplier to define the requirements for matching the motor and ASD to the load, including input transformer if required, to the power system such that they operate as a system with no unsatisfactory transient, torsional, heating, or power-quality problems. The ASD/motor/load system shall be suitable for the ship's service conditions, including the ambient temperature and humidity conditions; input power-supply voltage and minimum and maximum short-circuit capacities of the source; and auxiliary power-supply voltage.

Solid-state electronic power conversion equipment introduces electrical noise due to the frequency conversion process. Therefore, noise mitigating equipment such as power filters (passive band filters, active filters, ASDs fitted with AFEs, di/dt filters, braking resistors, and properly shielded cables) are vital to each system. This equipment must be identified as vital components for each and every vital system.

This equipment may also be used in a non-vital system; however, the nature of their operation may influence the entire power system. While failure mode and effects analysis (FMEA) deals with vital system and vital components, these devices may influence and must be treated as such. All requirements such as redundancy for single point failure must be analyzed for those components. Individual component health should be monitored and appropriate protective features be added to the control system so that the safety of the system can be properly maintained with no adverse effect to the other system.

2.2 Clause 23. EMI/EMC/RFI

"The power system and user loads should be designed and/or selected to ensure EMC. The specifications for the power system and loads should limit EMI and RFI. Applicable standards include MIL-STD-461, IEC 61000 series, and IEEE Std C63.12."

3 IEEE Std 45.8 –SYSTEM ENGINEERING Clause 5.5 Variable Frequency Drive Applications

"Cables for variable frequency drive (VFD) applications require special consideration because of large change in voltage with respect to time (dV/dt), harmonics, electromagnetic interference (EMI), reflected wave voltages, common mode currents, and potential induced voltages in adjacent cables and equipment.

The interconnecting cable between the motor and drive is an integral part of the entire system. Every VFD system is different and, typically has manufacturer's recommendations for cables and installation. Therefore consultation between the system designer, drive manufacturer, motor manufacturer, and the cable manufacturer is recommended for proper cable selection and installation. In the absence of specific manufacturer's recommendations refer to IEEE Std 1580 for recommended cable construction."

Appendix F   Steering Gear Motor Overload Protection Issue

1 Background

The steering gear motor overcurrent protection requirement is: The circuit breaker overload protection is with instantaneous trip only. The function of the instantaneous trip is to protect the motor from short circuit incident related fire or explosion.

The steering gear circuit breaker instantaneous trip setting recommendation in the IEEE-45-2002-35.5.2 *"at not less than 200% of the locked rotor current"*. This recommendation is considered ambiguous and should be revisited by IEEE-45 working group to make appropriate update.

USCG 46 CFR Subchapter 58.25-55 regulation is:

[The instantaneous trip should be set to "at least 175% and not more than 200% of the locked rotor current of one steering gear motor for an alternating current motor"]

2   Discussion

The circuit breaker instantaneous current setting must ensure that the circuit breaker does not trip during the motor starting to overcome locked rotor current requirement which may be within the instantaneous trip setting range. However, the instantaneous trip is to ensure the system protection to prevent equipment damage due to short circuit related fire or explosion

3   Recommendations

The steering gear circuit breaker instantaneous trip setting requirement needs to be further reviewed by IEEE-45.1 working group and if necessary the text be updated. Additionally the steering gear circuit breaker instantaneous trip setting must be coordinated during the short circuit analysis for proper protections such as (1) overcome the motor starting current and (2) and then trip setting at the earliest to limit arc energy.

Glossary

Ambient temperature Ambient temperature is the temperature of surrounding media such as air or fluid where equipment is operated or positioned.

ARC flash A type of electrical arcing fault, electrical explosion, or electrical discharge that results from a low impedance electric current path through air to a conductive plan or to another voltage phase in an electrical system.

Auxiliary Services System All support systems (e.g., fuel oil system, lubricating system, cooling water system, compressed air system, hydraulic system, etc.) that are required to run propulsion machinery and propulsors (ABS).

Azimuth thruster Rotatable mounting thruster device where the thrust can be directed to any desirable direction.

Bandwidth Generally, frequency range of system input over which the system will respond satisfactory to a command.

Circuit breaker frame (1) The circuit breaker housing that contains the current carrying components, the current sensing components, and the tripping and operating mechanism. (2) That portion of an interchangeable trip molded case circuit breaker remaining when the interchangeable trip unit is removed (100 AF, 400 AF, 800 AF, 1600 AF, etc.).

Continuous current rating (ampere rating) The designated RMS alternating or direct current in amperes which a device or assembly will carry continuously in free air without tripping or exceeding temperature limits.

Drawout circuit breaker An assembly of a circuit breaker and a supporting structure (cradle), so constructed that the circuit breaker is supported and can be moved to either the main circuit connected or disconnected position without removing connections or mounting supports.

VFD Challenges for Shipboard Electrical Power System Design, First Edition.
Mohammed M. Islam.
© 2020 by The Institute of Electrical and Electronics Engineers, Inc.
Published 2020 by John Wiley & Sons, Inc.

Electronic trip circuit breaker A circuit breaker that uses current sensors and electronic circuitry to sense, measure, and respond to current levels.

IDMT (Inverse Definite Minimum Time) Time/current graded overcurrent protection. Basically, the more current put through the relay, the faster it goes.

Instantaneous pickup The current level at which the circuit breaker will trip with no intentional time delay.

Instantaneous trip A qualifying term indicating that no delay is purposely introduced in the tripping action of the circuit breaker during short-circuit conditions.

Insulated case circuit breaker (ICCB) UL Standard 489 Listed non-fused molded case circuit breakers that utilize a two-step stored energy closing mechanism, electronic trip system, and drawout construction.

Interrupting rating The highest current at rated voltage available at the incoming terminals of the circuit breaker. When the circuit breaker can be used at more than one voltage, the interrupting rating will be shown on the circuit breaker for each voltage level. The interrupting rating of a circuit breaker must be equal to or greater than the available short-circuit current at the point at which the circuit breaker is applied to the system.

KAIC Kilo Amperes interrupting capacity. (Example: 65 KAIC or 65,000 AIC)

Long-time Ampere rating An adjustment that, in combination with the installed rating plug, establishes the continuous current rating of a full-function electronic trip circuit breaker.

Long-time delay The length of time the circuit breaker will carry a sustained overcurrent (greater than the long-time pickup) before initiating a trip signal.

Long-time pickup The current level at which the circuit breaker long-time delay function begins timing.

Making capacity (of a switching device) The value of prospective making current that a switching device is capable of making at stated voltage-prescribed conditions of use and behavior.

Molded case circuit breaker (MCCB) A circuit breaker assembled as an integral unit in a supportive and enclosed housing of insulating material, generally 20 to 3,000 A in size and used in systems up to 600 VAC and 500 VDC.

Peak let-through current The maximum peak current flowing in a circuit during an overcurrent condition.

Short-circuit delay (STD) The length of time the circuit breaker will carry a short circuit (current greater than the short-circuit pickup) before initiating a trip signal.

Short-circuit making capacity Making capacity for which prescribed conditions include a short circuit at the terminals of the switching device.

Short-circuit breaking capacity Breaking capacity for which prescribed conditions include a short circuit at the terminals of the switching device.

Time current curve (TCC) Method of ensuring selective coordination is to examine each overcurrent device's time-current curve (TCC) and verify for any value of current, that the protective device closest to the fault clears faster than any upstream device.

Clearing time The total time between the beginning of the overcurrent and the final opening of the circuit at rated voltage by an overcurrent protective device (ABS 4-9-1/5.1.15).

Constant horsepower range A range of motor operation where motor speed is greater than base rating of the motor, in the case of AC motor operation usually above 60 Hz where the voltage remains constant as the frequency is increased.

Constant torque range A speed range in which the motor is capable of delivering a constant torque, subject to motor thermal characteristics. This essentially is when the inverter/motor combination is operating at constant volts/Hz.

Constant Volts/Hertz (V/Hz) This relationship exist in AC drives where the output voltage is varied directly proportional to frequency. This type of operation is required to allow the motor produce constant rated torque as speed is varied.

Continuous rated machine The continuous rating of a rotating electrical machine is the rated kW load at which the machine can continuously operate without exceeding the steady state temperature rise (ABS) (4-9-1/5.1.2).

Converter The process of changing AC to DC, or AC to DC to AC. A machine or device for changing AC power into DC power (rectifier operation) or DC power into AC power (inverter operation).

Current-Source Converter A current source converter is characterized by a controlled DC current in the intermediate DC link. The line side network voltage is converted in a controlled DC current. A current source converter always uses an SCR bridge or an active front end to control the DC current.

Current limiting An electronic method of limiting the maximum current available to the motor. This is adjusted so that the motor's maximum current can be controlled. It can also be preset as a protective device to protect both the motor and control from extended overloads.

Deviation Difference between an instantaneous value of a controlled variable and the desired value of the controlled variable corresponding to the set point.

DOL starter Direct online starting system.

Drive A drive designed to provide easily operable means for speed adjustment of the motor within a specified speed range. The equipment used for converting electrical power into mechanical power suitable for the operation of a machine. A drive is a combination of a converter, motor, and any motor mounted auxiliary devices. Examples of motor mounted auxiliary devices are encoders, tachometers, thermal switches and detectors, air blowers, heaters, and vibration sensors.

Duty cycle The relationship between the operating and rest times or repeatable operation at different loads.

Efficiency Ratio of mechanical output to electrical input indicated by percent.

Engineering Control Center (ECC) ECC means a centralized engineering control, monitoring, and communications location (USCG).

Fail-safe Fail-safe means that upon failure or malfunction of a component or subsystem, the output automatically reverts to a predetermined design state of least critical consequence (USCG).

  A designed failure state that has the least critical consequence. A system or a machine is fail-safe when, upon the failure of a component or subsystem or its functions, the system or the machine automatically reverts to a designed state of least critical consequence (ABS).

Failure Mode and Effect Analysis (FMEA) A failure analysis methodology used during design to postulate every failure mode and the corresponding effect or consequences. Generally, the analysis is to begin by selecting the lowest level of interest (part, circuit, or module level). The various failure modes that can occur for each item at this level are identified and enumerated. The effect for each failure mode, taken singly and in turn, is to be interpreted as a failure mode for the next higher functional level. Successive interpretations will result in the identification of the effect at the highest function level, or the final consequence. A tabular format is normally used to record the results of such a study (ABS).

Fixed pitch propeller The propeller blades are fixed. There is no possibility of changing propeller pitch.

Induction motor An alternating current motor in which the primary winding on one member (usually stator) is connected to the power source. A secondary winding on the rotor carries the induced current (ABS 4-9-1/5.1.5).

Instrumentation A system designed to measure and display the state of a monitored parameter and which may include one of more of sensors, read-outs, displays, alarms, and means of signal transmission (ABS).

Inverter A converter in which the direction of power flow is predominately from the DC **Current Source Inverter (CSI)** An inverter in which the DC terminal is inductive and, as a consequence, the DC current is relatively slow to change. Modulation of the CSI acts to control the voltage at the AC terminal. The switches in a CSI must block either voltage polarity, but are only required to conduct current in one direction.

Voltage Source Inverter (VSI) An inverter in which the DC terminal is capacitive and, as a consequence, the DC voltage is relatively stiff. Modulation of the VSI acts to control the current at the AC terminal.

The switches in a VSI must block DC voltage, but be able to conduct current in either direction.

IPDE Integrated Product Data Environment, which features the capability to concurrently develop, update, and reuse data in electronic form.

IPT Integrated product team composed of representatives from appropriate disciplines working together to build successful programs, identify and resolve issues, and make sound and timely recommendations to facilitate decision-making.

Isochronous Constant speed and frequency irrespective of load.

LCI Type of adjustable speed current source drive called load commutating inverter.

Local control A device or array of devices located on or adjacent to a machine to enable it be operated within sight of the operator (ABS). Local control means operator control from a location where the equipment and its output can be directly manipulated or observed, e.g. at the switchboard, motor controller, propulsion engine, or other equipment (USCG).

Manual control Manual control means operation by direct or power-assisted operator intervention (USCG) (4-9-1/5.1.9).

Monitoring system A system designed to supervise the operational status of machinery or systems by means of instrumentation, which provides displays of operational parameters and alarms indicating abnormal operating conditions (ABS).

No-break power Power transfer from one source to another source instantaneously.

Non-periodic duty rated machine The non-periodic duty rating of a rotating electrical machine is the kW loads which the machine can operate continuously, for a specific period of time, or intermittently under the designed variations of the load and speed within the permissible operating range, respectively; and the temperature rise, measured when the machine has been run until it reaches a steady temperature condition, is not to exceed those given in 4-8-3/Table 4 (ABS).

Opening time (of mechanical switching device) Interval of time between the specified instant of initiation of the opening operation and the instant when the arcing contacts have separated in all poles. For circuit breakers: For a circuit breaker operating directly, the instant of initiation of the opening operation means the instant when the current increases to a degree big enough to cause the breaker to operate.

Overcurrent A condition that exists on an electrical circuit when the normal full load current is exceeded. The overcurrent conditions are overloads and short circuits.

Periodic duty rating machine The periodic duty rating of a rotating machine is the rated kW load at which the machine can operate repeatedly, for specified period (N) at the rated load followed by a specified period (R) of rest and de-energized state, without exceeding the temperature rise given in 4-8-3/Table 4; where N+R = 10 min, and the cyclic duty factor is given by N/(N + R) % (ABS) (4-8).

Podded propulsion A propulsion electric motor is installed in a watertight enclosure, where the motor is mounted on the same motor shaft.

PPE Personal protective equipment.

Propulsion machine A device (e.g., diesel engine, turbine, electric motor, etc.) that develops mechanical energy to drive a propulsor (ABS).

Propulsion machinery space Any space containing machinery or equipment forming part of the propulsion system (ABS).

Propulsion system A system designed to provide thrust to a vessel consisting of: one or more propulsion machines; one or more propulsors; all necessary auxiliaries, and associated control, alarm and safety systems (ABS).

Propulsor A device (e.g., propeller, or waterjet) that imparts force to a column of water in order to propel a vessel, together with any equipment necessary to transmit the power from the propulsion machinery to the device (e.g. shafting, gearing etc.) (ABS).

PWM PWM is a DC inverter system with the main power being rectified to produce DC and a self-commutated inverter to invert DC to AC to a variable frequency.

Redundancy Design Autonomous System An autonomous system is a system that can control and operate itself, independently of any control system or auxiliary systems not directly connected to it. Closed Bus: Closed bus often describes an operational configuration where all or most sections and all or most switchboards are connected together, that is, the bus-tie breakers between switchboards are closed. The alternative to closed bus is open bus, sometimes called split bus or split ring. Closed bus is also called joined bus, tied bus or closed-ring.

Redundancy Common Mode Failure A common mode failure occurs when events are not statistically independent, when one event causes multiple systems to fail.

Critical Redundancy: Equipment provided to support the worst case failure design intent.

Differentiation: Differentiation is a method to avoid common mode failures by introducing a change of producer of redundant systems based on the same principle.

Fail Safe Condition: The system is to return to a safe state in the case of a failure or malfunction.

Redundancy Hidden Failure a failure that is not immediately evident to operations and maintenance

Remote control A device or array of devices connected to a machine by mechanical, electrical, pneumatic, hydraulic or other means and by which the machine may be operated remote from, and not necessarily within sight of, the operator (ABS).

Remote control means nonlocal automatic or manual control (USCG) (ABS 4-9-1/5.1.8).

Remote control station A location fitted with means of remote control and monitoring (ABS).

RMS current The RMS is root-mean-square. The RMS current is the root-mean-square value of any periodic current.

Safety system An automatic control system designed to automatically lead machinery being controlled to a predetermined less critical condition in response to a fault which may endanger the machinery or the safety of personnel and which may develop too fast to allow manual intervention. To protect an operating machine in the event of a detected fault, the automatic control system may be designed to automatically (ABS):

slowdown the machine or to reduce its demand;

start a standby support service so that the machine may resume normal operation; or shutdown the machine.

Short-circuit breaking capacity Breaking capacity for which prescribed conditions include a short circuit at the terminals of the switching device

Short-circuit delay (STD) The length of time the circuit breaker will carry a short circuit (current greater than the short-circuit pickup) before initiating a trip signal.

Short-circuit making capacity Making capacity for which prescribed conditions include a short circuit at the terminals of the switching device.

Short time duty rating machine The short time rating of a rotating electrical machine is the rated kW load at which the machine can operate for a specified time period without exceeding the temperature rise given in 4-8-3/Table 4. A rest and de-energized period sufficient to reestablish the machine temperature to within 20°C (3.60°F) of the coolant prior to the next operation is to be allowed. At the beginning of the measurement the temperature of the machine is to be within 50°C (90°F) of the coolant (ABS).

Steering system A system designed to control the direction of a vessel, including the rudder, steering gear, etc. (ABS) (4-9-1/5.1.12).

Systems independence Systems are considered independent where they do not share components, such that a single failure in any one component in a system will not render the other systems inoperative (ABS).

THD Total harmonic distortion. It is accumulated distortion of fundamentals in a variable frequency drive.

Thyristor A component that conducts current in only one direction. Unlike a diode, the thyristor needs a firing pulse for it to start conducting current, after which it continues to conduct as long as there is current through it.

Time Current Curve (TCC) Method of ensuring selective coordination is to examine each overcurrent device's time-current curve (TCC) and verify for any value of current, that the protective device closest to the fault clears faster than any upstream device.

Tunnel thruster Produces fixed directional thrust (ABS 4-9-1/5.1.14).

Unmanned propulsion machinery space Propulsion machinery space that can be operated without continuous attendance by the crew locally in the machinery space and in the centralized control station (ABS 4-9-1/5.1.17).

Vital auxiliary pumps Vital auxiliary pumps are that directly related to and necessary for maintaining the operation of propulsion machinery. For diesel propulsion engines, the fuel oil pump, lubricating oil pump, and cooling water pumps are examples of vital auxiliary pumps (ABS).

Vital system or equipment Vital system or equipment is essential to the safety of the vessel, its passengers, and crew (USCG).

Bibliography

1 www.maib.gov.uk
2 IEEE-45-2002 Recommended Standard for Shipboard Electrical Installations
3 IEEE-519-1992 Recommended Practice and Requirements for Harmonic Control in Electric Power System
4 IEEE-519-2014 Recommended Practice and Requirements for Harmonic Control in Electric Power System
5 ABS Publication-150 Shipboard Harmonics
6 IEEE-45.1 Recommended Practice for Electrical Installations on Shipboard-Design
7 IEEE-45.2 Recommended Practice for Electrical Installations on Shipboard-Controls & Automation
8 IEEE-45.3 Recommended Practice for Electrical Installations on Shipboard-Systems Engineering
9 IEEE-45.5 Recommended Practice for Electrical Installations on Shipboard-Safety Considerations
10 IEEE-45.7 Recommended Practice for Electrical Installations on Shipboard-AC Switchboard
11 IEEE-45.8 Recommended Practice for Electrical Installations on Shipboard-Cable Systems
12 IEC 60092 Electrical Installations on Ships
13 IEC 61000 Series Electromagnetic Compatibility
14 IEEE-1566 Standard for Performance of Adjustable Speed Drives Rated 375 kW and Larger
15 IEEE-1584 Guide for Performing ARC Flash Hazard Calculations
16 IEEE-1662 Guide for the Design and Application of Power Electronics in Electrical Power System on Ships
17 Mil-STD-1399 US Navy Interface Standard-
18 NFPA-70 National Electric Code
19 NFPA-70E Standard for Electrical Safety in Work Places

VFD Challenges for Shipboard Electrical Power System Design, First Edition.
Mohammed M. Islam.
© 2019 by The Institute of Electrical and Electronics Engineers, Inc.
Published 2019 by John Wiley & Sons, Inc.

Index

VFD Challenges for Shipboard Electrical Power System Design, First Edition.
Mohammed M. Islam.
© 2020 by The Institute of Electrical and Electronics Engineers, Inc.
Published 2020 by John Wiley & Sons, Inc.